DO DOLPHINS EVER SLEEP?

DO DOLPHINS EVER SLEEP?

211 Questions and Answers
about Ships, the Sky and the Sea

Pierre-Yves and Sally BELY

SHERIDAN HOUSE

This edition first published 2007
by Sheridan House Inc.
145 Palisade Street
Dobbs Ferry, NY 10522
www.sheridanhouse.com

This book is an adaptation of *250 Réponses aux questions du marin curieux* by Pierre-Yves Bely, © 2004, Éditions du Gerfaut, Paris, France.

Every effort has been made to obtain permission to reproduce material in this book.

Library of Congress Cataloging-in-Publication Data

Bely, Pierre-Yves.

    [250 réponses aux questions du marin curieux] Do dolphins ever sleep? 211 questions and answers about ships, the sky and the sea / Pierre-Yves and Sally Bely. p. cm.
    " This edition first published 2007."

    ISBN-13: 978-1-57409-240-0 (pbk. : alk. paper)

    ISBN-10: 1-57409-240-5 (alk. paper)

    1. Navigation—Miscellanea. 2. Ocean—Miscellanea. I. Bely, Sally. II. Title.

    VK15.B25 2007

    623.8—dc22   2006030623

ISBN 13: 9781574092400
ISBN 10: 1574092405

Printed in China

Contents

Life in the Sea

The Sky

Wind and Weather

Ships

Yachting

Navigation

Life Aboard

Preface

The idea for this book came to me while crossing the Atlantic, west to east, by the long, easy southern route. So many weeks at sea had left me plenty of time to appreciate the changing color of the water, the beauty of the night sky during the graveyard shift, the flights of migrating birds, and to find myself prey to a growing, irresistible urge to turn off the GPS and pick up a sextant.

Some people love the sea for the sport it affords them, the chance to play with the wind and the waves. Others love it for its poetry, the fog-shrouded coasts and screaming gulls, adventurous winds and red sails in the sunsets. But the human animal is also curious by nature; we want to understand what we see and feel, and on a long passage, our curiosity has all the time in the world to develop. In my case, curiosity prompted questions about the sea, the sky, the subtleties of marine meteorology, the voyages of ancient navigators, the huge freighters passing by, the hydro- and aerodynamics of sailing ships, all that and so many other things that I wanted to know more about.

But after consulting my small, onboard library and talking with crewmembers and others in port, the answers I was able to come up with were often vague and unsatisfying, and I had to wait until I was home again to research the subjects and talk to specialists. And once I had the answers, it occurred to me that I might not be alone in my curiosity. So I wrote this book for all those of you who feel as I do, that understanding things is, in itself, a pleasure. I hope that, somewhere among these 211 questions and answers, you will find something to spark your curiosity and satisfy it, too.

The book was first published in French by Editions Gerfaut, and I particularly wish to thank their literary director, Pascal Pommier, for his encouragement and many useful suggestions regarding its original form. I also want to thank my wife, Sally, for the many hours she put into writing this English adaptation of the text, and Lothar and Janine Simon of Sheridan House for their enthusiastic support and editorial assistance in the publication of the English version.

Above all, I am grateful to the many scientists, engineers, historians and sailing professionals that I consulted with and who also helped by verifying the original manuscript. If this book has any claim to accuracy and thoroughness, it is thanks to their participation.

The complete list of my consultants appears in the French edition of this book and will not be repeated here, but I do want to extend my special thanks to Lucien Laubier, director of the Oceanographic Institute in Paris, Laure Fournier of Ifremer and the many scientists of that institution, my daughter Alexa Bely, professor of Biology at the University of Maryland, Michel Hontarrède of the French Weather Bureau, Eric Mas of Météo Consult, Hervé Garoche of the French Department of Lighthouses and Buoys, Pierre Frey of Bureau Véritas, Major Grant Walker of the U.S. Naval Academy, and naval architects Jean-Marie Finot, Pierre Gutelle and Guy Ribadeau Dumas.

Pierre-Yves Bely
College Park, MD
June 2006

Although the answers have been formulated with great care, certain inaccuracies and errors of interpretation may persist due to the large number of facts and explanations involved. I will, therefore, be grateful to readers who send in their suggestions for corrections, either by letter care of Sheridan House Inc, or by email to pbely@earthlink.net.

Notations :
Numbers between square brackets (i.e. [3]) apply to the list of references at the end of the book.

References to related questions are noted by the letter Q followed by the number of the question. For example, "Q.30" refers to question 30.

The Sea

1. How did the first seas form?

A photo taken from space shows something striking about Earth: our planet is blue.

View of Earth from space. Source NASA.

Oceans occupy 71% of its surface and their volume is enormous: 330 million cubic miles. Their average depth is 12,000 ft, five times the average height of the continents when measured from sea level.[1] Where did all this water come from?

A molecule of water consists of two atoms of hydrogen combined with one atom of oxygen (Q. 22). Hydrogen, the lightest chemical element, is found throughout the Universe. It is the fundamental substance that makes up the stars. Oxygen atoms, which are much heavier than hydrogen, are synthesized in novae and supernovae. These are stars whose existence ends in gigantic explosions in which oxygen as well as carbon, iron and other heavy elements are created. Our solar system (the Sun and the planets) formed 4.5 billion years ago from a cloud of "stardust" composed of the debris from such stellar explosions.

When gravity eventually caused our local dust cloud to become compressed enough, a new star running on thermonuclear energy, our Sun, was born. And the Sun then provided the heat energy needed to combine two atoms of hydrogen with one of oxygen to form water.

---

[1]Note that the oceans contain most of Earth's water: 97.2%. As for the rest, 2.1% is contained in glaciers, and only 0.7% is in rivers, lakes and aquifers. The atmosphere holds only 0.001%.

Water atoms were therefore present in the disk of debris left orbiting the Sun after its formation, and the planets, including ours, were formed from this matter. How did planet Earth manage to capture so much of this water? There are three current theories.

According to the first theory, water was simply one constituent of the dust and ice particles of the original disk orbiting the young Sun, the basic material that accreted to form the Earth. Then heat, produced by the gravitational compression of these accreted particles, caused the water to vaporize and be evacuated by a myriad of volcanoes; our primitive atmosphere was the result.

Five hundred million years later, the temperature of this first water vapor atmosphere dropped below the boiling point, and the vapor condensed to liquid water and fell to the surface in a deluge of warm rain. The water ran down to the lowest levels of the young terrestrial crust, accumulating into puddles, then into lakes, and finally forming seas and oceans.

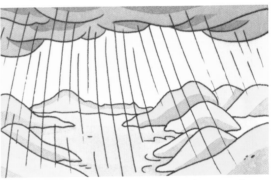

The oceans probably appeared after condensation of the water vapor that constituted the Earth's primitive atmosphere.

According to the second theory, the water came from the agglomerated dust and ice fragments (left over after formation of the planets) that abundantly bombarded Earth during its first billion years of existence.

According to the third theory, the water was brought to Earth by comets that bombarded it early in its history. Comets, too, are basically debris left over from the formation of the solar system and are rich in water ice.

Although we do not yet know which of these theories will prove to be correct, it is probable that all three sources contributed to the formation of our oceans. Volcanic emissions seem to have been the biggest contributor, however.

2. Why is Earth the only planet in our solar system with oceans?

Earth is singularly favored among the planets of our solar system: it rains on Earth, and there are rivers and oceans. The giant planets

(Jupiter, Saturn, Uranus and Neptune) have only very small solid cores surrounded by liquefied gas under intense pressure. No chance of finding continents or oceans there. The inner planets (Mars, Venus, Mercury) are rocky, more like Earth, but only Earth is lucky enough to have liquid water.

Of all the planets, only Earth has large quantities of water (sizes are to scale, but distances are not).

Mercury has hardly any atmosphere, and one side of it always faces the Sun. That side is so hot (800 °F) that lead would melt there.

Venus, our closest planetary neighbor, most resembles Earth in size and mass. This fact led to the belief that Venus was cloaked in vegetation and even inhabited. But 40 years of exploration by space probes have taught us that Venus is a lifeless desert without the slightest trace of water. It is closer to the Sun than we are, and if water ever was present there in the past, it evaporated long ago in the hellish temperatures that reign at its surface (900 °F).

Mars, being further from the Sun, is colder than Earth. In the 19th century it was thought that certain marks on its surface indicated canals, but we now know that this was an illusion. Still, space missions have revealed traces of dried river beds there, indicating that there was running water on the surface of Mars for a while after its formation. But volcanic activity stopped very early there and the water evaporated, leaving a dry, rarified atmosphere. If any water still remains near the surface, it can now only be in the form of ice buried in the soil, as temperatures at the surface are very low.

Our planet is thus particularly lucky [8]:

– we are just far enough away from the Sun for our water to be mostly in liquid form: 10% closer to the Sun and it would all have been vaporized, 10% farther away and it would all have been frozen.

- Earth is massive enough, and thus has enough gravity, to have retained its atmosphere and oceans (whereas most of the atmosphere of Mars, which is smaller, escaped into space),
- its axis of rotation is inclined just enough to produce seasons, which tempers the climate,
- and Jupiter is so positioned that its mass prevents comets and asteroids from bombarding us, which would threaten the existence of all life on Earth.

This exceptional set of circumstances is unlikely to occur very often around nearby stars, but that does not mean they will never occur anywhere. Our galaxy contains 100 billion stars, and there are more galaxies in the Universe than there are stars in our galaxy alone!

3. Does all the water that evaporates from the ocean eventually return to it?

Every day, 100 billion cubic meters of water evaporate into the atmosphere.[2] This water then condenses into clouds, falls as rain, and flows back to the oceans in rivers. In 37,000 years, a quantity of water equal to the entire amount contained in all the oceans thus goes into the atmosphere, then returns to Earth's surface[3].

A small amount of this water vapor does not return to Earth, however. In the upper atmosphere, solar radiation breaks some of it down into oxygen and hydrogen. The oxygen remains in the atmosphere but the hydrogen, a much lighter molecule, escapes into

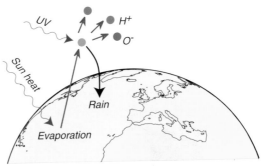

A little of the water vapor that evaporates from the oceans escapes into space.

space. Five gallons of water every second are thus permanently lost to us. That corresponds to a 10 foot drop in sea level since the oceans first formed.

---

[2]One third of the solar energy received by Earth is absorbed by the evaporation of water from the oceans.

[3]As for the water locked up in ice in the mountains and high latitudes, its confinement, too, is temporary, although the cycle is a much longer one. Depending on variations in Earth's climate and the successive phases of glaciations, this confinement lasts from a few thousand to a few million years.

4. Is there a difference between an ocean and a sea?

Strictly speaking, a sea is a body of salt water of secondary importance that is partly or completely surrounded by land. Most seas are connected to oceans by straits or channels.[4]

Seas (in blue) and oceans (in red).

The term "ocean" is reserved for those vast stretches of water that separate continents. It comes from the name of the Greek god Oceanus, the oldest of the Titans. His father, Uranus, was the lord of the heavens, and his mother, Gaia, was the goddess of Earth. He married Tethys, the goddess of the sea, and the Oceanid nymphs were their daughters.

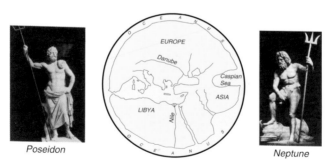

Poseidon   Neptune

Ancient map of the world (Hecataeus 500 BC.). The waters encircling the world were the realm of Oceanus. The god of the sea later became Poseidon for the Greeks and Neptune for the Romans.

---

[4]But there is a certain ambiguity between *lake* and *sea*. The difference is neither size (Lake Victoria is larger than the Aral Sea and the Dead Sea), nor salinity (the Caspian Sea has a salinity of 0.13%, while the Great Salt Lake is at 2%).

In Greek mythology, Oceanus was one of the "elder," first-generation gods; he represented the eternal flow of the world's oceans. The elder gods were eventually defeated in a great war by a new generation of gods who lived on Mount Olympus, and among these was Poseidon. This younger god of the sea was particularly associated with the Mediterranean. Neptune was originally the Roman god of fresh water, but he eventually merged with Poseidon to become the Roman god of the sea.

5. Why is the sea salty?

There are a lot of salts in seawater: 37 g per liter. If we could dry up the oceans and spread all the resulting salt over the Earth, it would make a layer 150 feet thick! Most of this is ordinary table salt, sodium chloride (NaCl), but there is also a little bicarbonate, magnesium, calcium and potassium in the form of chlorides and sulfates.

It would seem reasonable to believe that seawater is salty because running water leaches minerals out of the continental rocks, and that these minerals subsequently reach the sea by river transport. The story is actually more complicated. If the oceans simply received minerals washed out of the continents, they should become saltier and saltier. That is not the case. An analysis of sediments shows that the salinity of the seas has not changed over the past 200 million years.

Another fact contradicts this overly simple view. As shown in the table at right, the proportions of the various salts in river water and seawater are different. Seawater is essentially ordinary table salt in solution, while river water contains mostly calcium and bicarbonates.

Relative concentration of salts (%) [80]

Element	Seawater	Rivers
Chlorine (Cl^-)	55.0	5.7
Sodium (Na^+)	30.6	5.8
Sulfate (SO_4^{--})	7.7	12.1
Magnesium (Mg^{++})	3.7	3.4
Calcium (Ca^{++})	1.2	20.4
Potassium (K^+)	1.1	2.1
Bicarbonate (HCO_3^-)	0.4	35.1
Other substances	0.3	15.4

So the oceans do not simply receive material eroded from the continents. The different concentrations of solutes in seawater are actually the result of an *equilibrium* between the original concentrations in the primordial seas, the solutes brought in by the rivers, and what is lost through exchange with the sea beds and the atmosphere by biological transformation or through "precipitation" (sedimentation), when concentrations exceed the carrying capacity of water.[5]

And each chemical substance has a different history. The sodium (Na) is brought in by rivers in relatively small quantities, but it dissolves readily in water (hence its high concentration) and is evacuated through sedimentation. Potassium is adsorbed by the clays at the bottom of the seas. Calcium is taken up by living organisms to construct their shells or skeletons. When these organisms die, the calcium falls to the bottom of the sea and becomes part of the sediment.

But chlorine (Cl) takes no part in any of the geological cycles. Unlike calcium, sodium, potassium and manganese, it is not trapped in sediments or rocks and can only remain in the seawater. Moreover, rivers bring in only minute quantities of this element. Thus it appears that the chlorine now found in seawater must have been a constituent of Earth's primitive atmosphere and must have become dissolved in water as soon as the original atmospheric water vapor became liquid (Q. 1). And it has simply remained there ever since.

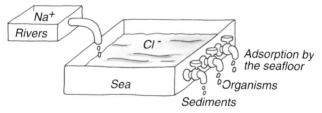

As in those classic "running spigot" problems in school math books, the concentrations of the various elements in sea water are the differences between what the rivers bring in and what diverse processes remove...except for chlorine, which was present in seawater from the beginning and is not removed by any geological or biological process.

So as it turns out, the salt in seawater is not just ordinary salt dissolved in the sea, as one might expect. The chlorine and sodium there have completely independent origins and "life cycles." It is only when seawater is allowed to dry up that these two elements unite to form salt crystals, making us think that the sea contained salt all along.

---

[5]The rate of this natural recycling is expressed as "residence time," the time it takes for an element to be completely evacuated, one way or another. The residence time for sodium is about 120 million years; for calcium, 1 million years; and for chlorine it is almost infinite. The residence time for water itself is relatively short: 37,000 years.

6. What explains the presence of islands out in the middle of the ocean?

We find it easy to explain the presence of islands near a coast: they are broken off fragments of the continent, or they look as if they had broken off because the sea level rose in an already eroded landscape. But an island in mid ocean? How can we explain Bermuda, the Azores, the Falklands, Hawaii and other Polynesian islands located in water that is thousands of feet deep?

The answer here is that such islands are volcanic, originating in "hot spots" in the Earth's crust. A hot spot is an area in the upper mantle connected to a deep pocket of lava. Such pockets exist under all the tectonic plates, continental as well as oceanic.

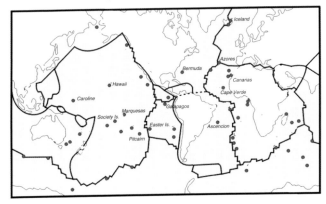

The principal tectonic plates and oceanic hot spots (red dots)

Lava can continue to flow from a hot spot for millions of years. As the ocean-bottom plate moves over the stationary hot spot, the lava emerges at different places on the sea bed, forming a series of volcanoes that can eventually create a string of islands.[6] This would explain the origin of the Hawaiian archipelago which includes the major islands plus a series of reefs extending all the way to Midway, and would also explain the major archipelagos of Polynesia.

Several hot spots are located along mid-ocean ridges (areas where two tectonic plates are currently diverging). The magma plumes they produce add to the volcanic activity that normally occurs in such areas and increase its volume. Iceland, the Azores, and Ascension Island were formed this way.

Most oceanic volcanoes remain below sea level. As each becomes extinct, it leaves behind an underwater mountain with a conical summit, called a *seamount*.

---

[6]This theory is somewhat controversial. Although a stationary hot spot does seem to explain the origin of the Hawaiian Islands, certain other hot spots appear to "drift."

A volcano that emerges, however, creates an island that eventually erodes in the air and is left with a flat summit. Such islands then begin to sink as the seabed

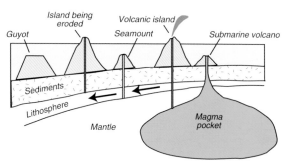

beneath them slumps under the weight of accumulated sediment.[7] These islands then disappear under the waves, creating underwater mountains with truncated summits called *guyots* (named after Arnold Guyot, a 19th century American geologist of Swiss origin). There are over 10,000 seamounts and guyots in the Pacific.

7. How fast are the continents drifting apart?

The continents float on the Earth's inner layer of magma (the *asthenosphere*) as though on the surface of a viscous liquid. They drift along, propelled by the great convection currents that slowly churn our planet's upper mantle. Some 250 million years ago they were all grouped together in one mega-continent, Pangaea[8] that began breaking up 50 million years later to create the Atlantic Ocean, among other features. Today, North America and Europe are still moving apart at the rate of about 2 inches per year.

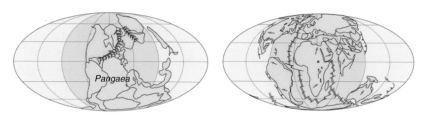

At left, Pangaea, and, at right, the position of the continents 50 million years ago.

---

[7]Sediment layers on the sea floor can become several kilometers thick in 100 million years.

[8]There was another, much earlier, Pangaea 550 million years ago.

8. Is there any truth to the myth of Atlantis?

According to the ancient legend, Atlantis was a large island beyond the Pillars of Hercules (Gibraltar) inhabited by a civilization that was very advanced for its time. Around 9600 BC, its armies had conquered the whole world except for Greece. Suddenly, the island was swallowed up by the sea, disappearing without a trace and thereby sparing Greece the humiliation of being conquered.

The legend of Atlantis (which may have given its name to the Atlantic Ocean) is one of the most persistent ever. More than 5000 books and articles have been published on the subject proposing all kinds of sites for the "lost civilization": an island in the mid Atlantic, the island of Bimini in the Caribbean, Antarctica, etc. The earliest reference we have for it is in Plato, and it is quite possible that Plato himself invented the whole thing in order to satirize Athens.

But if the legend has any basis in fact, the volcanic eruption on the island of Santorini (also called Thera) near the island of Crete could have been at its origin. This eruption, which occurred in 1650 BC, was one of the biggest to occur in the past 10,000 years.[9] About 8 cubic miles of

rock exploded, creating a column of gas and ash 20 miles high that covered Turkey and the island of Crete with ash, and causing a giant tsunami. Whether or not this event gave rise to the myth of Atlantis, the cataclysm it produced was probably responsible for the collapse of the Minoan civilization on Crete [28, 16, 17, 57].

9. Is the surface of the sea perfectly round?

More precisely, if we disregard wave heights, tidal bulges, and the slight variations in sea level caused by currents, is the sea's surface spherical?

The ocean at rest has a particular shape called a *geoid*. It is approximately the ellipsoid that would be formed by a homogenous liquid turning around the axis of its poles in 24 hours, as the Earth does.

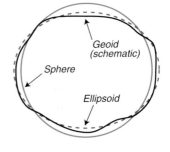

---

[9]Another eruption of approximately the same magnitude took place in 1883 at the Krakatoa, Indonesia, creating a tsunami 130 feet high and killing 36,000 people.

However, many factors cause the geoid to differ from this theoretical ellipsoid. The most important factor is local differences in gravity due to the different densities of rock in the Earth's crust, and of all the matter inside it, down to the very center of Earth. This effect causes surprisingly large variations in sea level: 160 meters (about 500 ft)! The lowest sea level occurs off the coast of India, and one of the highest levels is off the coast of New Guinea.

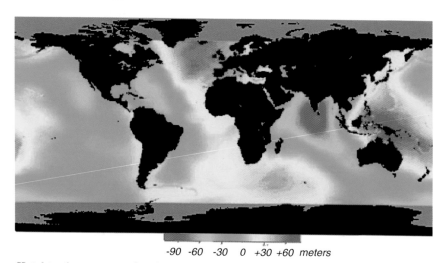

-90 -60 -30 0 +30 +60 *meters*

Height of mean sea level.   Data obtained by radar from the NASA and CNES TOPEX/Poseidon satellite.

Small local variations in the height of the sea floor add to these large-scale gravitational effects. An underwater mountain is denser than water and locally increases gravitational attraction, while the presence of a marine trench will diminish it.

Contrary to what you might expect, though, the increased attraction of gravity due to an underwater mountain creates a "hill," not a "valley" at the surface. The sketch at right shows why: the surface of the water remains perpendicular to the pull of

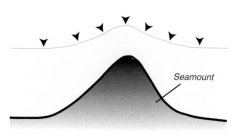

Seamount

gravity. This surface distortion is not negligible: a 6,000 foot underwater mountain creates a hill of water about 30 feet high on the surface above it.

To these gravitational effects we must add variations in height due to temperature differences in the surface waters (warm water is less dense, so water levels rise) and to variations in atmospheric pressure (a high-pressure zone causes the sea level to drop, while a low-pressure zone makes it rise 10 cm for every 10 millibars of difference in pressure). These effects are fairly weak, causing variations of 3 feet or less.

So even if we do disregard the variations due to the mid-ocean tides, currents and waves, the surface of the sea is far from being a simple sphere.

10. Is mean sea level rising measurably?

About 20,000 years ago, at the end of the last period of glaciation, the mean sea level was about 400 feet below the current level. When the Earth's temperature increased, the glaciers melted and the sea level rose at the rate of 0.4 inch per year to reach its present level about 8000 years ago. That level remained essentially the same until very recently.

The mean sea level has risen by about 5 inches over the last century, however, probably because the temperature of the atmosphere is increasing due to the greenhouse effect. And even if we were to slow down or completely stop the production of greenhouse gasses today, the sea level would continue to rise for some time, as the ocean reacts very slowly to climate changes thanks to its huge thermal inertia. If all the Earth's icecaps and glaciers did finally melt, our mean sea level would rise by about 200 feet.

11. How does wind create waves?

Wind does create waves, obviously. Just blow on your cupful of coffee and you will produce the same effect. But if you stop to think about it, you might wonder why the wind doesn't simply slide over the water, leaving its surface flat.

(a)   (b)   (c)   (d)

First we have to remember that all surface perturbations in water create waves. When water is at rest, its surface is the result of an equilibrium between gravity attracting it downward and buoyancy (water pressure — think Archimedes) pushing it upward. If you throw a stone into water, the stone pushes the water down (a), then, under the effect of buoyancy, the water moves up again (b), reestablishing the equilibrium. But inertia will carry the moving mass of water higher than the original surface (c). When the mass of water drops again under the effect of gravity, it descends too far once more, and so on and on. This opposition between buoyancy pushing upward and gravity pulling down would continue indefinitely if the internal friction of the water (viscosity) did not eventually bring it to a stop.

Why should air create this kind of disturbance at the water's surface? Because the air itself is turbulent. When wind blows over an initially calm, smooth surface, the water scarcely moves because it is much denser and more vis-

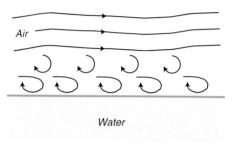

cous than air, but it does slow the air's movement. Slowed down at its base, the air then "summersaults" to form eddies. So the air layer in contact with the water surface becomes turbulent. The turbulent layer creates localized high- and low-pressure areas on the water which perturb its surface and create ripples, just as we saw above. The difference here is that, in addition to the force of gravity, another force is present in the ripples that also combats buoyancy: surface tension.

Surface tension, a force produced at the surface of a liquid, works at the molecular level. It is surface tension that pulls the water you spray onto your smooth, shiny car hood into rounded drops, and that makes rain, too, fall in round drops.

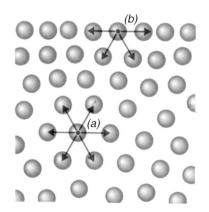

The reason for this is depicted in the sketch at left. The molecules inside a volume of water (a) are attracted by other water molecules in all directions (Q. 22) and are in overall equilibrium , while the molecules at the surface (b) are only attracted by those below it. The surface then acts like an extremely thin membrane under tension, and the water tends to be pulled into the configuration that gives the smallest surface area. For a small quantity of water, this force will pull the water into a spherical shape because, for any given volume, a sphere is the shape with the smallest surface area. If the mass of water is large and its surface is perturbed, this force will tend to flatten out the surface.

This effect is only important for small amplitudes, hence for short wavelengths. Surface tension becomes negligible and gravity becomes dominant when the wavelength approaches two centimeters.

Initially, wind creates small, scarcely visible waves of very short wavelengths.[10] At this scale, where surface tension is stronger than gravity, they are called "capillary waves." They are the first small waves that form under a gust of wind.

Once the surface is slightly perturbed by air turbulence and these small waves have formed, the surface disturbance is amplified by the aerodynamic effect. The wind, which by then is itself perturbed by the small waves, whirls around behind them, making them grow bigger.

Once the wavelets created by the passage of air over water have appeared, the air begins to whirl around behind their crests and the phenomenon is amplified.

This process produces waves of all wavelengths in response to the extremely variable speed and direction of the gusting wind. The amplitudes slowly increase until the waves break. This occurs when the angle at the summit of a wave falls below 120 degrees, which corresponds to a height of 1/7th of the wavelength. The crest is too steep then and, propelled by the wind, it breaks, forming whitecaps.

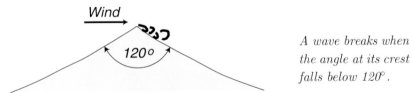

A wave breaks when the angle at its crest falls below 120°.

Waves with short wavelengths break first, transferring part of their energy to the larger ones that overtake them. The small waves thus disappear bit by bit to the benefit of longer ones that may be higher and store more energy.

This process continues until the energy furnished by the wind is completely absorbed by the movement of the water particles inside the waves (Q. 14).

---

[10]These wavelets tend to have a wavelength of 1.7 cm, the wavelength corresponding to the minimum speed of waves on water.

12. What is the difference between sea and swell?

When speaking of the state of the sea, waves created locally by wind action are called "sea." These are short, chaotic waves of various periods that break when the wind is strong.

The wind's effect on waves depends on the wind's speed, duration and the distance over which it blows unobstructed, a distance called the *fetch*. On the open sea, the fetch is usually determined by the size of the atmospheric low that produces the wind.

Waves travel beyond the fetch and are slowly transformed into regular waves with long wavelengths, all having the same direction. This is called the "swell." This uniformity in direction and wavelength takes place thanks to a filtering effect caused by the fact that short waves travel more slowly and die more quickly than long ones. These long waves, when unobstructed, can travel for hundreds or even thousands of miles.

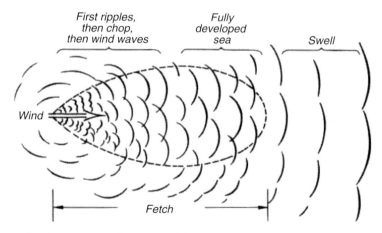

Wave size increases in the zone of fetch up to a maximum value which depends on the extent of the fetch and the speed of the wind. Waves propagate beyond the fetch, are filtered, and produce the swell.

Since there is always a wind somewhere on Earth, the sea is never completely calm. A global view is given by the following figures.

Beyond the subtropical belt, wave heights and wind speeds vary a great deal with the seasons, particularly around Antarctica where there are no continents to block the winds and currents that circulate around the globe (Q. 89). Winds there are particularly strong in the austral winter but become more moderate in summer, opening passageways for adventurous navigators.

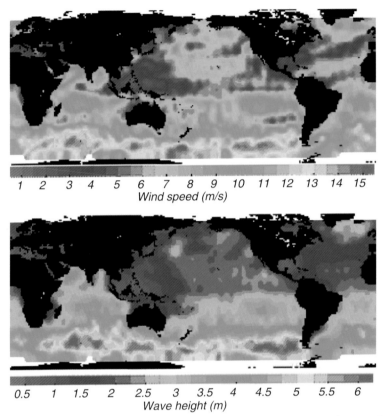

Wind speed and significant wave height during summer in the northern hemisphere. Naturally, the stronger the wind, the larger the waves. The smallest waves (dark blue) are thus found mainly in tropical and subtropical oceans where winds are relatively weak. (Data obtained by radar measurements taken by the Franco-American satellite Topex in July 1998 — Source: Aviso/CNES [2].)

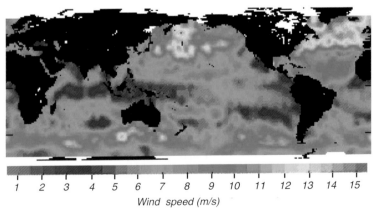

Typical wind speeds during the austral summer. Data for January 1994. Source: Aviso/CNES.

13. What is the true geometric shape of a wave?

We tend to think of waves as sinusoidal, and that is certainly true of light and sound waves and also the shape of oscillating strings in musical instruments. But for ocean waves, only swells are roughly sinusoidal in shape. The approximate shape of a wind wave in a fully developed sea is a geometric figure known as a *trochoid*. This is the curve described by a point *inside* a wheel turning on a flat surface.

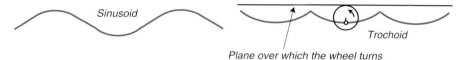

The sinusoid has a rounded crest, while the trochoid has an angular crest.

At sea, of course, it is rare to find a series of perfectly identical waves. The state of the sea at any given moment is the sum of the waves coming in from different directions and with different amplitudes, all due to different successive wind conditions that may have prevailed quite far away.

The superimposition of waves coming in from different directions tosses the boat around and can be dangerous when the waves are big.

This is what created the dramatic situation during the 1979 Fastnet, when waves coming in from the southwest, the south and the northwest converged under storm conditions, creating chaotic seas with whitecaps.

General sea conditions are created by the superimposition of waves with different directions, amplitudes and periods. The periods are usually between 5 and 15 seconds.

14. How fast do waves travel in water?

Whether created by wind or a tidal wave (tsunami), a wave's speed depends on its own wavelength and the depth of the water.

In deep water, the water particles acquire a circular motion as the wave passes, with their direction of rotation at the crest being the same as the direction of the wave, and in the opposite direction in the trough. The viscosity of the water causes the particles at the surface to drag those just beneath them along, so that this circular movement is repeated deeper down but with ever smaller circles, since friction between the particles absorbs some of their energy. The circular movement becomes negligible at a depth equal to one half the wavelength.

When water depth is less than half a wavelength, the wave "feels the bottom." Friction against the bottom then slows the movement of the water particles and several things happen. The wave slows down, shortens (its wavelength decreases), becomes asymmetrical (the slope in front of the wave progressively steepens because, as the water becomes shallower, this is the first part of the wave to shorten) and the motion of the particles becomes elliptical. In very shallow water the elliptical paths become so flat that the water particles simply move back and forth.

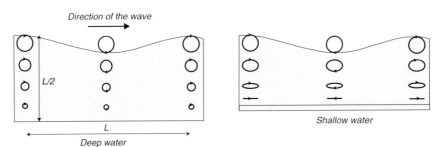

The path of water particles is circular in deep water (deep with respect to the wavelength) but becomes elliptical in shallow water.

In deep water, the speed of the wave is given by the formula [11]

$$c = \sqrt{\frac{g\lambda}{2\pi}} \quad \text{or, equivalently} \quad c = \frac{gT}{2\pi},$$

where c is the speed, g, the acceleration of gravity, λ, the wavelength and T, the period ($T = \lambda/c$).

---

[11] This formula is not easy to derive. Wave theory was formulated in the 19th century by a number of famous physicists (Froude, Stokes, Rankine, Rayleigh) and calls for rather complex mathematics. It is very well confirmed experimentally.

Expressed in knots, the speed is approximately equal to 3 times the period expressed in seconds and to 1.34 times the square root of the wavelength expressed in feet:

$$c \text{ (in knots)} \simeq 3 \times T \text{ (in seconds)} \simeq 1.34 \sqrt{\lambda} \, (\lambda \text{ in feet}),$$

(or $2.43 \sqrt{\lambda}$ for λ in meters), which also gives:

$$\lambda \text{ (wavelength in feet)} \simeq 5.1 \, T^2 \, (T \text{ in seconds}).$$

We see that speed increases with wavelength (or with the period: $T = \lambda/c$): a long swell moves faster than a little wave. We also see that speed is independent of a wave's height.

The table below gives a few examples of wave speeds in the open sea as a function of their periods or wavelengths. The period of a swell is usually between 7 and 12 seconds. The longest period ever timed was 22.5 seconds, corresponding to a wavelength of 2500 feet, or 0.4 nautical mile.

Speed and wavelength of a wave as a function of its period

Period, T (seconds)	1	2	4	6	8	10	12	16
Wavelength, λ (feet)	5	20	82	184	328	512	737	1310
Speed, c (knots)	3.0	6.1	12.1	8.2	24	30	36	48

In shallow water, the speed of a wave is given by

$$c = \sqrt{gh} = 3.4 \sqrt{h} \text{ (in feet)}$$

where h is the depth of the water. We see that, in this case, speed is no longer a function of the wavelength and depends only on water depth. This case of a wave in shallow water naturally applies to waves approaching a coast, but also applies to waves with very long wavelengths compared to the water depth, such as tidal bulges and tsunamis.

Tsunamis (also called tidal waves) are solitary waves created by earthquakes occurring underwater or near coasts, and by underwater volcanic eruptions. Their wavelengths are usually longer than 60 miles. Since oceans are only a few miles deep, tsunamis must therefore be treated like waves in shallow water, and their speed is proportional to the depth of the ocean. If we use 13,000 feet as an average depth, the speed of a tsunami will be about 380 knots, almost as fast as a commercial airplane.

Authorities thus have only a few hours to react and evacuate endangered inhabited zones. Since the Hawaiian islands and other islands in the Pacific are so vulnerable to tsunamis originating in the highly seismic zones of the Pacific rim (Chili, Japan, Alaska, California), they have established an automatic warning system based on seismographic data.

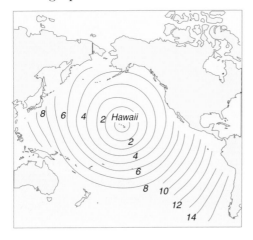

Time in hours required for a tsunami from the Pacific rim to reach the Hawaiian islands.

Out at sea, a tsunami presents no danger at all; it would simply consist of a wave about 3 feet high with a very flat crest extending over dozens of miles. On a boat you would not even notice that you have been lifted up. When a tsunami reaches a coast, however, its wavelength shortens and its amplitude increases, just as for regular waves. Then such a wave can reach a height of 100 feet and cause enormous damage.

15. Why do big waves often come in threes?

In a fully developed sea, waves are chaotic. You cannot expect the slightest regularity [3]. But beyond the fetch zone, or if the blow has lasted long enough, waves become filtered (Q. 12) and have essentially the same direction and similar wavelengths. In such cases, the waves combine in a regular way, sometimes reinforcing each other, sometimes canceling each other, to create what physicists and musicians call a *beat*.

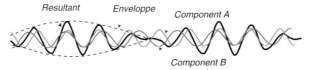

Waves with similar periods combine, resulting in a wave whose amplitude (the envelope on this graph) varies regularly. The highest wave is preceded and followed by waves of comparable height.

The concept of the "wave train," although an idealized situation, is fairly representative of reality. If you encounter one big wave, you can expect to encounter another even bigger one, followed by yet another, almost as big.

An added consequence of beat is that strong waves do tend to return at regular intervals. The old saying to the effect that, if you encounter one big wave, "another will come along 7 waves later," though not to be taken literally, is not just pure myth, either [79].

16. How big was the biggest wave ever recorded?

It is hard to judge the height of a wave from the deck of a boat. Many have claimed to have sighted waves 60 to more than 100 feet high, but the highest one ever measured with certainty was 110 feet [3]. The sighting took place in the middle of the Pacific during a typhoon.

Wave measured from a US Navy vessel, the RAMAPO, *480 feet long, caught in an extraordinary storm in the North Pacific in 1933. The officer on watch on the bridge saw the crest of the wave aligned with the crow's nest, and from this was able to calculate the height of the wave: 110 feet.*

Such gigantic waves represent the extreme among the large waves that form during long-lived storms. But enormous, very steep "solitary" waves have also occasionally been spotted in relatively calm seas. Isolated waves occurring in such conditions are called *rogue waves* or *freak waves*.

Several hypotheses have been proposed to explain the origins of rogue waves, such as abrupt changes in the topography of the seabed (the water encounters a steep continental shelf or a submarine mountain, for example) or the presence of strong currents.

One of the most dangerous regions in the world from this point of view is at the southern tip of Africa, in the waters of the swift Agulhas current. Since the current is stronger in the middle than at its edges, swells born in Antarctic storms that encounter it grow crescent-shaped and converge towards its center, creating huge, pyramid-shaped waves. These very steep waves have destroyed many a ship.

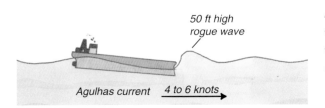

50 ft high
rogue wave

Schematic view of a very steep solitary wave in the Agulhas current.

Agulhas current   4 to 6 knots

17. Can you really calm storm waves by pouring oil on them?

Throwing oil overboard to calm a stormy sea is a venerable part of marine folklore, the last desperate defense measure. The idea was probably born when whalers reported that the waves diminished when whale oil was spilled on them — and the effect is real enough. Waves are reduced, for example, when an oil tanker accident leaves the sea covered with an oil slick. The thin layer of ice floating on the sea when it is about to freeze over has a similar effect.

This effect is due to the surface tension of the covering film. When a wave passes, it stretches the film, but the energy expended is not recovered when the film regains its shape afterwards. That energy is dissipated as heat. As a result of this energy loss, the formation of wavelets by wind action is inhibited, and waves cannot form without wavelets (Q. 11). During one experiment in a lake covered by a film of detergent (which has the same effect as oil), waves were negligible under 35-knot winds, whereas they were about a foot high in detergent-free waters [79].

Now, if waves cannot form on a film-covered surface, the calming effect is minimal for waves originating elsewhere, unless the film is extremely thick (a large fuel oil slick, for example). Nevertheless, a film does tend to keep waves from breaking: whitecaps and sea spray disappear. Unfortunately, this effect can only calm moderate-sized waves. When the weather is really bad, the dissipation of energy due to surface tension no longer counts when measured against the enormous power of breakers.

When all is said and done, even if you do happen to have a few barrels of oil aboard, you cannot hope to calm a furious sea via this old practice.

18. Why do waves grow larger when they reach a beach?

A wave is strongly affected by the sea floor when water depth drops to less than half the wave's wavelength, and this modifies the wave's mode of travel. Its period remains the same but its speed is reduced (Q. 14). Since wavelength is equal to the speed of a wave multiplied by its period, the wavelength is also reduced, and the amplitude (height) increases.

As a wave approaches a beach, friction against the bottom causes the trough to slow down more than the crest, creating an imbalance. If the slope of the bottom is gentle, this effect is weak and the crest "spills over" as it advances (sketch a: spilling breaker). If the slope is somewhat steeper, the crest loses its equilibrium, plunges over the trough and "breaks" (sketch b: plunging breaker). In even steeper conditions, the slope of the bottom counteracts the slope of the wave's flank, the wave advances practically intact and surges on the shore without breaking (sketch c: surging breaker).

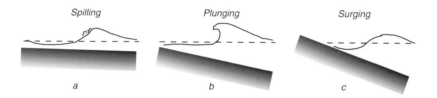

Spilling	Plunging	Surging
a	b	c

Waves breaking near a coastline, including the breakers sometimes found at harbor entrances, are dangerous for boats.

When you cannot avoid going into such an area, it is imperative to keep the boat facing into the crests of the waves. The engine must be kept at full power to maintain some speed over the moving water of the wave, or the rudder will become useless. A boat caught sideways in a breaking wave is in danger, as the weight of the breaking water and the lateral resistance of the boat's hull and keel create a moment (torque) that can make it capsize.

Sailors may justly fear large breakers, but surfers adore them: the upward movement of the water in front of the crest allows surfers to maintain their positions on the wave as it travels forward.

Surfers can be compared to someone going down an "up" escalator. The water of the wave moves up under them, compensating for their weight and keeping them just below the crest

19. Why is the Bay of Biscay so dangerous?

Is it true that the Bay of Biscay is often "gray and tempest tossed," or is that just another one of those popular myths, like the one about the Mediterranean always being blue and placid?

The truth is that fear inspired by the Bay of Biscay is entirely justified. Winter storms are fierce and frequent there; the weather changes abruptly as cold fronts associated with lows pass through, and the seas, which are often crossed, can be extremely uncomfortable. The abrupt rise of the sea floor to a depth of only 110 meters, due to the continental shelf, causes the reinforcement of all waves with periods of over 12 seconds.

Moreover, the Rochebonne Plateau, the local shoal 30 miles offshore at a depth of only about 30 feet, brings additional risks. No doubt about it, the Bay of Biscay's unpleasant reputation is richly deserved.

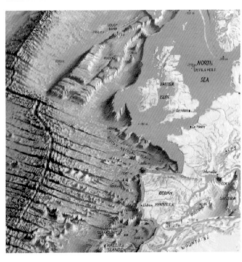

The Atlantic seafloor off the coast of Europe. Note the abrupt change in depth on the floor of the Bay of Biscay.

20. Is there any difference between sea ice and the ice in an iceberg?

Icebergs are pieces of polar glaciers that break off and float out to sea. Although their name implies that they are made of ice (frozen water), they are really composed of compacted snow. Sea ice is ice that forms on the surface of the sea itself.

Seawater, which contains 37 g of salts per liter, freezes at -1.9 °C (29 °F). During the freezing process the salt is expelled and the ice that forms is practically pure water. This allows seafarers from the upper latitudes to use it as ice cubes in their drinks without affecting the taste.

Freezing starts with the formation of small ice crystals about 1 millimeter long at the surface of the sea, giving it an oily appearance. The crystals grow, eventually fusing to form ice plaques from 1 to 10 feet in diameter and about 4 inches thick. At this stage it is called *pancake ice.*

Sea ice.

Photo: Michael Van Woert, NOAA NES-DIS, ORA.

Pushed around by the wind and swell, these ice blocks bump into each other and stick together, forming ice floes with rough, uneven surfaces. Sea ice covers a large percentage of the polar waters in winter. In the summer it melts, partially in the Arctic and almost completely in the Antarctic.

21. Ice is transparent but icebergs are white. Why?

Ice that forms when quiet water freezes is indeed transparent. You can see this on the edge of a lake or in the ice cube tray in your freezer.

In such solid blocks or masses of ice, the water molecules are tightly bound in a crystal lattice, and light photons cannot easily penetrate their electron envelopes: they simply pass between the molecules, unaffected, and the ice is transparent (Q. 183).

But icebergs are not solid blocks. They are made of accumulated, compacted snow, and in snow the many tiny facets of the individual snow crystals reflect

Ice cubes are transparent, while icebergs are white.
Photo: NOAA.

most of the sunlight, scattering it at all wavelengths (i.e. all colors) in random directions (Q. 66). The scattered sunlight that then reaches our eyes therefore looks white and opaque.

22. Matter is generally denser as a solid than as a liquid; so why do icebergs float?

Be it an ice chip in your soda or an iceberg big enough to sink the TITANIC, ice blocks of all sizes and shapes float on water. Now it is true that iron, lead, and most other substances are heavier as solids than as liquids, and that makes water an exception. Why?

As you know, water is made up of one atom of oxygen bound to two atoms of hydrogen (H_2O). When atoms come together to make a molecule, it is electrical attraction that keeps them together (what is called *valence* in chemistry), and normally the result of such a union is elec

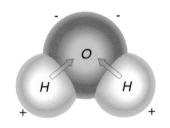

trically neutral. What is different about the water molecule is that it is not neutral *locally*. The lone electron of the hydrogen atom is attracted by the atom of oxygen, leaving its proton "naked" on the other side. This creates a strongly positive pole. And since the oxygen atom has now become over-endowed with electrons, its "back" forms a slightly negative pole.

This means that the water molecule is bipolar and tends to be attracted to neighboring water molecules, with the naked hydrogen side of one molecule cozying up to the well endowed side of the oxygen atom in another molecule. This is called "hydrogen bonding."

In the gas state, thermal energy keeps the water molecules so agitated that they are constantly colliding and rebounding. That keeps them spaced well apart, overall.

In the solid state of water (ice), a lower temperature means less molecular agitation: the molecules orient themselves into a rigid crystalline structure and are glued into place by hydrogen bonds. But the combination of attractions and repulsions between molecules causes them to keep a certain distance from their neighbors, leaving small empty spaces, tiny holes, in the crystal lattice.

In the liquid state, however, where kinetic (thermal) energy is higher, the molecules are free to orient themselves. Here, hydrogen bonding pulls them snugly together. And since these bonds are constantly forming and reforming, the molecules end up, on average, closer together than they are when inside ice crystals.

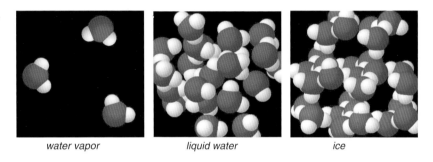

water vapor   liquid water   ice

This situation does not occur in most substances because most molecules are not dipolar. Usually, electrical forces do not keep the molecules in solids apart, so they crowd together.

It is thanks to this unique feature of the water molecule that Earth's oceans remain liquid in winter. If ice were denser than liquid water, it would form at the surface in contact with the cold air, then would immediately sink to the bottom, where it would tend to build up rather than melt. Little by little our oceans would have filled up with ice. No marine life would exist. Luckily for us, ice does float, stays at the surface, and insulates the liquid water beneath it.

The polarity of the water molecule is responsible for another important property: it is an excellent solvent for substances that are not electrically neutral, especially those that dissociate into ions, such as the salts found in seawater. The molecules of water are electrically attracted to the ions, turning their oxygen side towards 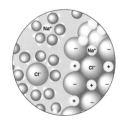 positive ions (such as Na^+) and their hydrogen side towards negative ions (such as Cl$^-$). The water molecules thus tend to surround the ions, separating them; the salt is "dissolved."

23. If the Sun attracts the Earth much more than the Moon does, why does the Moon create larger tides than the Sun?

The Moon is the main cause of our tides. This is true of the Atlantic and Pacific coasts of North America and, indeed, of most of the coasts on Earth.

Yet the Sun's attraction is certainly stronger than the Moon's. We revolve around the Sun, not around the Moon. The Sun, though 390 times more distant than the Moon, is 27 million times more massive. According to Newton's law, the attraction between two bodies is equal to the product of their masses divided by the square of their distance. The attraction of the Sun is therefore $27 \cdot 10^6/390^2 = 178$ times stronger than that of the Moon.

Still, we do not fall into the Sun, and the reason is that its attraction is compensated by centrifugal force acting on the Earth as it revolves around the Sun.

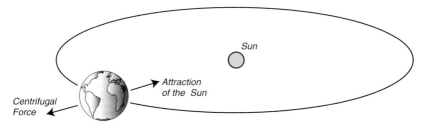

The attraction of the Sun is compensated by centrifugal force acting on the Earth during its annual revolution around the Sun. The solar tide is not due to the Sun's attraction per se, but to the differences in the strength of its attraction at various points on Earth. The same is true for the lunar tide.

So what counts is not the attraction itself but the *difference* of attraction as a function of the distance to the Sun (or the Moon). For example, a point on the Earth's surface facing the Sun will be slightly more attracted than a point on the opposite side of the Earth. Instead of varying as the *square* of the distance (as attraction does), the differential attraction varies as the *cube* of the distance,[12] so that the tide caused by the Sun is about half of that due to the Moon

---

[12] The derivative of $1/d^2$ is $-2/d^3$.

$(27 \cdot 10^6/390^3 = 0.45)$. Overall, then, 2/3 of the tide is due to the Moon and 1/3 to the Sun.

Now, although this is true in general, the presence of continents and local coastal configurations can diminish or amplify tidal effects, so that in certain places the tidal component due to the Moon will overwhelm that of the Sun, or, conversely, become negligible.

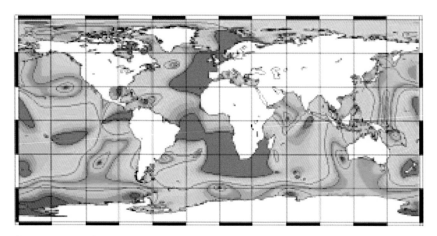

Comparison of the tidal components due to the Sun and the Moon. The zones in blue and green are those where the diurnal component, mainly due to the Sun, dominates. Red, brown and yellow indicate zones where semidiurnal tides, mainly due to the Moon, predominate. Source: Legos/CLS.

24. If the tide is mainly due to the attraction of the Moon, why is there also a high tide on the opposite side of Earth, the side facing away from the Moon?

Since the tide is primarily due to the attraction of the Moon, one would expect that the waters of the oceans would be drawn to the side of Earth that faces the Moon, as in the figure on the right. Then there would only be one tide a day, not the usual two.

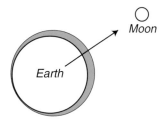

What actually happens is that the Moon does not turn exactly around the center of the Earth, but around the center of gravity of the Earth-Moon couple. The same is also true of the Earth going around the Sun: it is the center of gravity of the two bodies combined, Earth + Moon, that revolves around the Sun, while the Earth describes little undulations on either side of the perfect orbit.

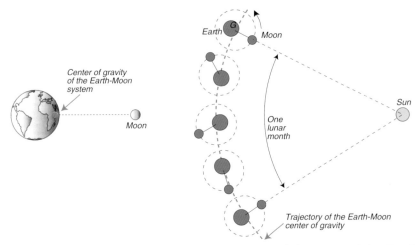

In one lunar month, the Moon does not turn around the center of the Earth, but around the center of gravity of the Earth-Moon system, and the Earth does the same.

Since the Earth is much heavier than the Moon, the center of gravity of the couple Earth-Moon is located near the Earth's center, about 1,700 km below the Earth's surface.[13] In one lunar month, the Earth thus revolves around this point and, like all revolving bodies, it is subjected to centrifugal force. Nonetheless, contrary to what one might expect, this centrifugal force does not depend on the distance to the center of rotation because we are looking at it in a non-rotating reference frame.

When you twirl an object tied to a string, the centrifugal force is stronger on the outer side of the object, which is further from you, than on the inner side, which is closer. But that is not the case with Earth because the Earth is not phys-

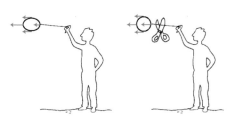

ically attached to the Earth+Moon's center of gravity around which it revolves. It is *at every instant* in the situation of an object that has been twirling on a string, but whose string has just been cut. Such an object would be in "free fall," with all its parts moving together.

---

[13]The mass of the Earth is 81 times that of the Moon, so the center of gravity of the couple Earth-Moon is at 1/80 of the distance between Earth and the Moon. Since the distance from Earth to the Moon is 384,000 km, the center of gravity of the Earth-Moon couple is 4,700 km (384 000/81) from the center of the Earth, which is about 1,700 km below the surface of the Earth.

The mutual attraction between Earth and the Moon plays the part of the string maintaining the Earth on its little orbit around the center of gravity of Earth+Moon. But the centrifugal force is the same everywhere on and inside the Earth, and is equal to the force acting at the center of gravity.

At the center of Earth, centrifugal force exactly counteracts the attraction of the Moon on the entire Earth (the distance between the Moon and the Earth has adjusted itself so as to create this equilibrium). But this is not true for other points on Earth, for the attraction of the Moon varies with the distance to it. This attraction is stronger on the side of Earth where the Moon happens to be, and less strong on Earth's opposite side.

Attraction from the Moon   *Centrifugal force*

Two forces combine to create the tides: gravitational attraction of the Moon (left) and centrifugal force due to the Earth's rotation around the Earth+Moon center of gravity (right).

So, ultimately, lunar attraction overcomes the centrifugal force on the side of Earth facing the Moon and this creates the predictable 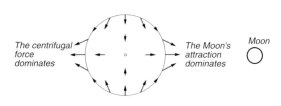 bulge of water. On Earth's other side, it is the centrifugal force that overcomes the lunar attraction and creates the second tide [10, 53].

25. Why are tides so much bigger in some places than in others?

There is about a 3-foot tide in the middle of the Atlantic. Why should there be 10 to 15 foot tides in the English Channel and tides of over 45 feet in the Bay of Fundy, in Canada?

This is easy to understand if we compare the sea to a little tub filled with water. If you move the tub gently, once, the water will surge to one side, and then oscillate back and forth with a certain period. The period will depend on the size of the tub: the larger the tub, the longer the period (a bowl has a shorter period than a bathtub). If we now shake the tub rhythmically with a period that is almost the same as the one that the water first responded with, the natural period for that tub, the oscillatory movement is amplified and the water will slosh out over the sides. This is the phenomenon of *resonance*: the amplitude is greatest when the exciting force has the same period as the natural period of the system being excited.

The Atlantic has a natural period of about 12 hours,[14] which is close to the period of excitation due to the Moon's pull (12 h 25 mn). This explains the rather big tides on the North Atlantic coasts.

In certain places, local topography enhances the general resonance effects felt in the Atlantic. The English Channel and, even more strikingly, the Bay of Fundy behave like funnels into which the sea rushes,

If you shake a tub of water at a certain frequency, it spills over.

and resonance is greatly amplified. In the Bay of Fundy, this creates the biggest tides in the world.

26. Why are tides so small in the Mediterranean?

The Mediterranean Sea is very deep (up to 15000 feet), but the Strait of Gibraltar, its connection to the Atlantic, is narrow and shallow (depth of less than 1,000 feet). This keeps the Mediterranean from having much contact with Atlantic tides, unlike the English Channel which is very open to the Atlantic and receives its tidal bulge head on. Tides in the Mediterranean are thus only affected by the pull of the Moon and the Sun inside the sea's own basin.

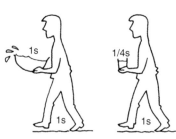

The analogy of the tub of water that we used in question 25 is useful again here. Water in a small tub has a period of about one second. That is pretty close to the period we impart to it as we walk with it, so it will be hard to keep it from sloshing out. However, if we walk around holding a glass of water, with an oscillation period of approximately 1/4 s, we have no sloshing problem.

[14]Since the Atlantic has an average depth of about 12,000 feet (4,000 m), a long wave travels at a speed of $\sqrt{gh} = 200$ m/s (Q. 14) that is to say 720 km/h. Since the distance between France and Canada is about 4,400 km, the natural period of the Atlantic at this latitude will be about 12 hours ($2 \times 4400/720$).

In the Mediterranean there are actually two large basins separated by a sill in the Strait of Sicily, between Sicily and Tunisia, which is at a depth of only about -1,100 feet. The resonance period of each of the basins (to the east and west of Sicily) is only 2 to 3 hours.[15] This is much shorter than the periods of the two tidal forces (Moon:12h, Sun:24h). So not only is there no amplification in the Mediterranean, but the effects of the Moon and the Sun are actually damped. As a result, the tides there are usually under 20 cm (8 inches).

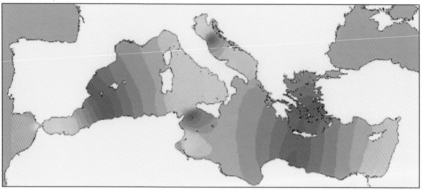

Amplitude of the lunar component of Mediterranean tides. The amplitude is 150 cm for red, 50 cm for green, 20 cm for light blue and 0 for dark blue. Source: Legos/CNES (CEFMO model).

This same phenomenon of very small tides is also found in the Gulf of Mexico and the Baltic Sea.

Because Mediterranean tides are so minimal, marinas can use the so-called "Med mooring." In this type of docking practice, the boat is tied up perpendicular to the dock and an anchor or mooring holds the bow. Floating docks are not needed. This means that more room is available along the piers, and a simple plank is all one needs to disembark.

---

[15]Since each of the Mediterranean basins has an average depth of about 2,000 meters, the speed of a tidal bore there will be about $\sqrt{gh} = 140$ m/s, or 500 km/h (Q. 14). Since the basins are both about 1,200 km long, their fundamental period is about $1200/500 \simeq 2.5$ hour.

27. What explains the almost total absence of tides in Tahiti?

The tidal bulge cannot really move around the Earth in as un-complicated a way as indicated in question 24 because the continents block its progress. The bulge actually travels in a complex manner, often in a circular path, around points with zero tidal effects called *amphidromic points*. When a basin of water is moved to and fro, it creates a wave that goes back and forth; the water level goes up and down at either end, but the depth in the middle of the basin remains constant. If we move the basin in a circular path, the crest of the wave we create will begin to rotate around a central point where the water level stays steady. This is what happens in the oceans. Due to the rotation of the Earth, the tidal bulge circles around a number of amphidromic points rather than simply moving from east to west. Rotation is generally clockwise in the Southern Hemisphere and counterclockwise in the Northern Hemisphere.

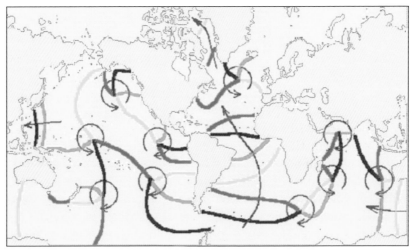

Position of the crest of the tide at three-hour intervals: red, yellow, green and blue at 0, 3, 6 and 9 hours respectively. The arrows indicate the direction of rotation around the amphidromic points.

Each component of the tide has its own circulation around different, often nearby, amphidromic points. At neap tide in Tahiti, the semi-diurnal bulges due to the Moon and Sun have the same amplitude but opposite phases, meaning that the one is at its maximum below mean water level when the other is also at its maximum, but above mean water level. The change in water level cancels out and the tide becomes imperceptible. In the spring, when the two tidal bulges are additive, the tide is at its highest but is still quite small, barely 30 cm. As it happens, the tide occurs around noon and midnight then, whence the name *solar tide* that is sometimes given to the tide in Tahiti.

28. What causes the large surges of water that sometimes occur even in well protected harbors?

It can be surprising: the open sea is relatively calm, the harbor well protected by breakwaters, and yet the water inside is sloshing back and forth hard enough to snap your mooring lines. This oscillation, called a *seiche* by oceanographers, is due to stationary waves inside the basin being excited by the open-sea swell when its period is just right. It is a resonance effect similar to what happens inside a wind instrument, where a very strong tone is produced even though the air column is only mildly excited by the vibrations of the reed.

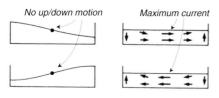

To explain this, let us consider what happens when sloshing the water back and forth in a basin. The water level stays unchanged in the middle of the basin (this is a node), while it goes up and down at either end (these are the crests and troughs of the oscillations, or *antinodes*). The horizontal movement (the current) in the water is zero at each end and maximum in the middle. In this oscillation, the length of the basin, L, corresponds to one half the wavelength of the wave, λ. If the water depth is small compared to the length of the basin, we have the conditions governing shallow water, and by applying the formula for the speed of a wave (Q. 14) we get $T = 2L/\sqrt{gd}$, where T is the period of the oscillation and d is the water depth in the basin.

The same reasoning applies to a harbor, but here the basin is open at one end. We still have an antinode at the end of the harbor, but the node is now at the harbor entrance, since the water level there is determined by the sea. The length of the basin therefore corresponds to $\lambda/4$, and the resonance period is given by: $T = 4L/\sqrt{gd}$.

The fundamental period of oscillation varies from less than a minute for small harbors to dozens of minutes for long, deep harbors. The exciting force can have any of several different origins, a sudden change in atmospheric pressure, for example, or a storm or an earthquake.

The effects can be devastating. The fishing fleet in the port of Ciudadella in Minorca was heavily damaged by a seiche with a period of 10 minutes and an amplitude of 2 m [25]. Port Tudy, a small harbor on the island of Groix, located a few miles off the coast of Southern Brittany, is occasionally subjected to seiches due to a wave with a period of 5 minutes that occurs between the island and the European continent. The amplitude of this wave can exceed 3 feet and cause violent currents inside the harbor.

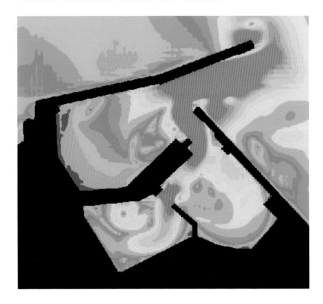

Seiche current in Port Tudy (red: 1.5 m/s, yellow: 0.7 m/s, green 0.2m/s). Source: ACRI, Simulation by David Lajoie.

Note that the seiche is a phenomenon similar to that of a tide. In essence, tide is an ocean-wide seiche excited by the gravitational forces of the Sun and the Moon.

29. Why are cold seas green and warm seas blue?

In spite of what you probably learned at school, water is not a colorless substance. It can look colorless in a bottle, but when seen through a great thickness, it is blue. You can prove this to yourself by diving to the bottom of a swimming pool and looking up. The blue color of pure water is due to the absorption and scattering of light by the water molecules (Q. 66). The effect varies with the wavelength of light: red is the first color to disappear, while blue persists into deep waters.

Tropical seas with very pure water are blue, but that is generally not true of the Atlantic in temperate latitudes. The water there is rather green. This is because temperatures in the higher latitudes are lower, favoring an abundance of phytoplankton (Q. 38). And phytoplankton contains chlorophyll, a substance that absorbs blue and red light but transmits the green.

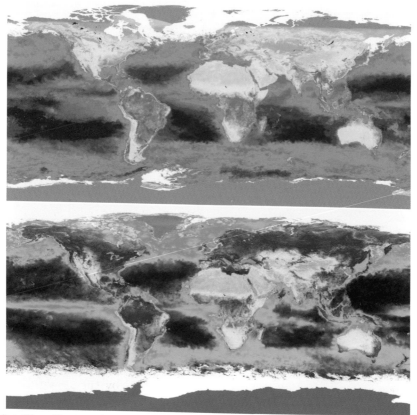

Satellite (SeaWifs) view of the color of the sea in winter (top) and in summer (bottom) of the Northern Hemisphere. The green regions are areas with high concentrations of phytoplankton. The blue areas, with low phytoplankton productivity, are the oceans' "deserts." Source: NASA.

30. What causes phosphorescence in the sea?

Phosphorescence at night in breaking waves, bow waves, or the ripples from an oar can be surprising.

Phosphorescence in the sea is created by swarms of microscopic plants, specifically an alga, the noctiluca, in the group of algae called dinoflagellates. This alga, which makes up part of the plankton, becomes luminous when agitated. It is a defense mechanism. When the water is stirred up, the alga "imagines"[16] that it could be caused by

---

[16]Just a metaphor, of course. Darwinian evolution produced this reaction.

A particularly impressive example of bioluminescence at Vieques near Puerto Rico.   Courtesy of Frank Borges LLosa (frankly.com).

a tiny crustacean, like a copepod, that feeds on phytoplankton. So it flashes a little light that might attract a fish that might gobble up the crustacean. These flashes of light are very brief, on the order of tenths of a second, but since a large number of noctilucas are emitting them, the phosphorescence seems to last a good while.

The flash of light is the result of a chemical reaction. Dinoflagellates contain a protein, luciferin, that emits light when oxidized by contact with the air brought in by the agitated water.[17] Note that, strictly speaking, the term phosphorescence (or fluorescence) should not be used here, as it refers to light that is *re-emitted* by a body after being absorbed by it. The proper term is *bioluminescence*, the light produced by a biochemical reaction.

31. What causes the foam on whitecaps?

Sea foam is just like the foam on dishwater that contains a few drops of detergent or the foam on a cappuccino: it consists of bubbles of gas enveloped by liquid membranes.

When a liquid is agitated, air mixes in with it. But foam never forms on pure fresh water, even if it is shaken or stirred violently. Whitecaps never form on lakes, no matter how strong the wind is.

---

[17]Luciferin is also responsible for phosphorescence in fireflies, certain bacteria, and many invertebrate animals.

The reason for this is that any air bubbles introduced into fresh water almost immediately cluster together and the liquid membranes separating them disappear.

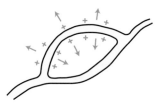

A bubble is maintained by electrostatic action.

You cannot make foam by rubbing your hands in fresh water, but use a little soap and the bubbles appear. The role of soap is to provide electrical charges (of the same sign, positive or negative) on the two surfaces of liquid membranes. These electrical charges produce repulsive forces that prevent the liquid membrane from becoming too thin and thus keep the bubble from bursting too soon.

The superficial layers of the ocean contain all kinds of organic matter, including plankton, that serve as "wetting agents," substances that act as detergents.

By the way, when a foam bubble on a whitecap bursts, it sprays salt into the air. It is now believed that most of the salt carried about by our atmosphere comes from burst sea foam bubbles.

32. Why is the wet sand on a beach darker in color than the dry sand?

Dry sand is a mixture of air and solid grains that are more or less transparent (quartz, for instance), whereas wet sand is a mixture of these same grains and water. The difference in appearance stems from the different indexes of refraction for water and air.

A ray of light passing from one medium into another is bent, or refracted, and the amount of bending increases as the difference between the two indexes of refraction increases. The refractive index of quartz is about 1.5. If we replace the air (index 1) by water (index 1.33), a ray of light going into a grain of sand will be less bent. The path of a ray of light as it goes through many sand grains will thus be much more direct in wet sand than in dry sand. As a result, the light penetrates more deeply into wet sand and tends to come back up towards the observer much less: the wet sand thus looks dark. But in the case of dry sand, the light rays are more strongly bent, their paths zigzag this way and that through the quartz grains, and in most cases they return towards the observer, making the sand look lighter.

Refraction effects in dry sand (left) and wet sand (right). Since light rays are refracted less in wet sand, a lot of them are lost into it and it appears dark.

33. Why is the South Pacific also called the "South Seas"?

The expression "The South Seas" evokes the tropical dreamworld of Polynesia — we think of Stevenson's *In the South Seas* — but it originally referred to the entire Pacific Ocean.

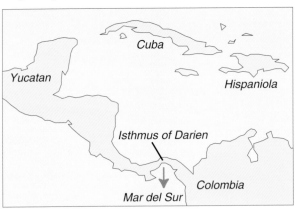

In 1513 Vasco Nuñez de Balboa, the Spanish explorer, crossed the Isthmus of Panama (later the site of the Panama Canal) on foot, coming from the Caribbean [64]. On reaching the far shore of the isthmus, which runs east-west, Balboa saw a huge body of water to the south that he named the *Mar del Sur*, or South Sea. And after that, the Spanish world quite naturally began referring to the Atlantic Ocean as the *Mar del Norte*, the North Sea.

Seven years after Balboa's discovery, Ferdinand Magellan's search for a southwest passage linking the North Sea and the South Sea was crowned with success when he discovered the straits that now bear his name. He was so happy to enter the calm waters of the vast "southern" ocean after his difficult passage past Cape Horn that he renamed the South Sea the "Pacific."

34. What exactly is the Gulf Stream?

The Gulf Stream is a strong, warm, surface current created by the convergence of the current from the Gulf of Mexico and that of the Caribbean. It crosses the Atlantic and arrives, much diminished, off the coast of Norway, where it is recycled by the great oceanic "conveyor belt" (Q. 35). At its strongest, off the coast of United States, it is about 50 miles wide and has a speed of 3 to 5 knots. It remains strong as far as southern Newfoundland before slowly losing its strength.

The Spanish navigators knew of the Gulf Stream's existence; Ponce de Leon had reported it off the coast of Florida as early as 1513. In the 17th century, when English colonies began developing on the American continent, the captains of American merchant ships and whalers were very familiar with the current. But strangely enough, their English homologues seemed not to know about it.

At left, the first map of the Gulf Stream, drawn by Benjamin Franklin in 1769. At right, a map showing the mean surface temperature of the ocean in 1996, as obtained by satellite. The Gulf Stream is the red band with a temperature of about 28° C (green areas are at 18° C, blue ones are at 10° C).   Source: NOAA/NASA.

The first map of the Gulf Stream was drawn by Benjamin Franklin. At the time, the administrators of the English mail service had noticed that colonial American mail boats (the *packet ships*) often covered the distance between America and England in two weeks' less time than their English counterparts. They questioned Benjamin Franklin, then director of the colonial American postal service, about the discrepancy, but he had no idea how to explain it. Franklin then brought up the subject with a cousin of his, a man who captained a whaling ship and who was quite aware of the Gulf Stream's existence. All the colonial captains knew about the current, he told Franklin. When they sailed for England, they would position themselves in it, then would avoid it on their way back. Aided by his cousin, Franklin proceeded to draw up the first map of the Gulf Stream and gave it to the English postal services. But the English captains completely discounted this fantastic tale of an "oceanic river" and continued to follow their accustomed route for nearly another hundred years [63].

35. Are the deepest ocean waters stagnant?

Local tidal currents and the great surface currents like the Gulf Stream and the Humboldt current have been known for a long time, but until fairly recently, it was generally believed that the ocean depths were stagnant. The fact is that slow but powerful currents churn through all the ocean waters at all depths.

The driving force behind this circulation is differences in water density in different places. This is the same phenomenon that causes water to circulate naturally, without a pump, in a central heating system: the water in the hot water furnace, which is less dense, rises from the cellar, goes into the radiators, cools down during its passage through the radiators, becoming denser, then drops back down to the furnace. In the sea, there is no hot water furnace to heat the water, but there are two refrigerators to cool it: the Arctic in the north and Antarctica in the south. Let us first take a look at the latter. Antarctica, which arrived at the South Pole over 200 million years ago via continental drift, has served as the collecting point for a considerable amount of ice ever since and has become the coldest place on our planet. Thanks to the same principle as that in central heating, but in reverse, the waters of the Antarctic become cold in contact with this frigid continent and so they dive, forming deep currents that then penetrate into the Indian and Pacific Oceans.[18]

These deep currents collide with the continents, are carried back up to the surface by upwellings, warm up as they circulate towards the south, and collect in the South Atlantic. From there, they begin moving north, receive a little push as they cross the zone of trade winds, and mix in with the waters of the Gulf of Mexico to form the Gulf Stream.

After crossing the Atlantic, one portion of the Gulf Stream turns south again and is recycled in the equatorial circulation system, but the other portion goes north, towards Norway. Chilled in the Arctic and aided by an increase in salinity (acquired by reabsorbing the

---

[18]The effects of salinity differences must be added to the purely thermal effects. Just as cold water is denser and tends to sink, so saltier water is denser and tends to do the same. In scientific terms, circulation due to both of these factors is termed *thermohaline* (thermo = temperature, haline = salt).

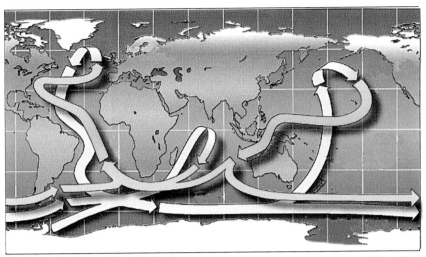

The great oceanic conveyor belt. In red, the surface currents, and in blue, the deep-sea currents.   *Courtesy of* la Recherche.

salt expelled from the surface water as it freezes), this portion dives deep and moves towards the Antarctic, thus completing the cycle. When taken all together, these cyclic currents form a gigantic oceanic "conveyor belt" of planetary dimensions.

The sheets of water in these currents move along very slowly, from a few millimeters to a few centimeters per second. The rate of flow is enormous, however: 20 to 30 million m³/s (25 times the rate of flow of the Amazon River, or the equivalent of all the rain that is falling at every instant on Earth). More than 90% of the water in all the oceans participates in this circulation, which thus gently stirs the whole of it. But all this takes time: a full circuit requires about 1500 years.

This conveyor belt that transports cold water from the polar regions towards the warm southern zones, and vice versa, participates in the thermal exchange between the poles and the tropics, just as atmospheric circulation does (Q. 88). It is this exchange that keeps the poles from becoming too cold and the tropics from being too hot. There are indications that the conveyor belt stopped working at certain times in the past, a "seizing up" of the system that may be either the cause or a consequence of Earth's glaciations.

36. Is the Bermuda Triangle's bad reputation justified?

The Bermuda Tri-
angle, a region of the
Atlantic delimited by
Florida, Bermuda and
Puerto Rico, has been
the site of a number
of disappearances af-
fecting both ships and
planes. And it is true
that some of these dis-

appearances have been unexplained, such as the one in 1945 off the
coast of Florida involving a squadron of American Avenger bombers,
or the loss without a trace of the American naval vessel USS Cyclops
with 300 personnel aboard in 1918.

Building on such more or less well confirmed disappearances and
other occasional odd events, sensation-seeking journalists and writers
with fertile imaginations began clamoring that the Bermuda Triangle
was under the control of occult forces, even going so far as to suggest
that the disappearances were the work of an advanced civilization
living under the sea! Then the movies seized on the misadventures
of the bomber squadron in the film *Close Encounters of the Third
Kind*.

The U.S. Navy and Coast Guard, which can be expected to know
the seas pretty well, do not consider the recorded disappearances
any more mysterious than many other unexplained accidents. In the
case of the squadron, for instance, a navigational error could have
led to its loss due to a fuel shortage. It should also be noted that
this region is peculiar in having a zero magnetic variation which,
if overlooked, can cause an error of about 10 degrees. In seas like
these, with their numerous reefs, this amount of error could easily
prove fatal. The Gulf Stream is another possible factor: it is a rapid,
turbulent current that can speedily erase all proof of a disaster. And
finally, human error is not to be minimized; these tropical seas, so
inoffensive looking, are crisscrossed by many pleasure craft whose
crews are not always able to handle the unexpected weather changes
and numerous navigational dangers of the region.

In fact, when these so-called mysterious disappearances are carefully analyzed, it becomes clear that the reports often contain more
fiction than fact. A librarian at an American university who examined the sources used by the sensation-seeking writers was able to
prove that most of the disappearances they wrote about are not as

strange as all that [40]. For example, some have claimed quite dis-
honestly that the disappearance of a certain vessel occurred in calm
seas, when official documents mention a big storm. Or again, "a boat
disappeared mysteriously" but in fact wreckage of it had been found
and the cause of the shipwreck clearly explained.

Perhaps the most persuasive proof that this region is not under
the control of evil forces is that the Lloyd's of London archives of
accidents at sea indicate that the Bermuda Triangle is not a particularly dangerous sector of the ocean, no more than all the rest of it,
at any rate.

37. What laws regulate the freedom of the seas?

The control of coastal waters has always been a subject of rivalry
among world powers. And yet, seamen have also always considered
themselves to be members of a large family that transcended nations.
You came to the rescue of a ship in distress no matter what flag it
flew. There was a tacit agreement that the seas were no one's private
property.

It was a Dutch jurist, Hugo Grotius, who first expressed the principle of the free use of the sea by all countries with the publication
of his *Mare Liberum* (The Free Sea) in 1609. But somewhat later
another Dutch jurist proposed limiting this notion to the high seas,
and extending the sovereignty of countries with seacoasts to a strip
of water along their shores. In 1703 it was decided that the width of
such territorial waters would be equal to the reach of a cannon ball,
or about 3 miles. This principle was adopted by all countries, even
though some of them sometimes adopted a distance of 6 or even 12
miles.

Beginning in about 1930, countries slowly began to understand
the value of the continental shelf for oil, minerals, and fishing, and
some of them decided to extend their territorial waters. In 1945, at
the instigation of oil companies, President Harry Truman unilater-
ally annexed all of the continental shelf adjoining the United States
(this shelf extends out for about 200 miles). That move pushed other
countries to standardize the extent of their own territorial waters
and the question was brought before the United Nations in 1958. After long negotiations, a treaty was devised in 1982 at Montego Bay,
Jamaica, specifying that:

- the territorial waters, the zone over which a coastal nation has
  full sovereignty, extended for 12 miles, and that

— every coastal nation also had the exclusive right of exploitation of the sea and sea floor in a 200 mile-wide zone called the "exclusive economic zone" (EEZ).

This treaty went into effect in 1996, with the ratification of the 120th country [78]. The European countries have all ratified it but the United States has yet to do so.

With the 200 mile definition, EEZs occupy an enormous part of the seas. If you put all the EEZs together the area would represent a third of all the oceans, leaving only 2/3 as pure "high seas."

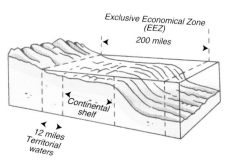

The Montego Bay treaty also affirms the right of all to circulate freely in international straits. In the Mediterranean, only the Strait of Gibraltar is subject to this convention. The others straits there (Dardanelles, Bosporus, Suez) are each under the jurisdiction of a single state. In these cases, the freedom to move through them is governed by regional agreements.

The freedom of the seas is governed by international law. The activities of navigation, maritime transportation, fishing, scientific research and aerial overflight are free to all. But in order to protect our increasingly fragile environment, these activities may have to be regulated in the future.

Life in the Sea

38. What is plankton?

Living organisms that float passively in the sea and are carried around by the currents are collectively called plankton. The term comes from the Greek *plankton*, meaning "which floats." Since the many different types of individual organisms constituting plankton are generally small and transparent, they are not usually visible even though they comprise 95% of all living matter in the sea. There are two types of plankton: the plant-like phytoplankton and the animal form called zooplankton.

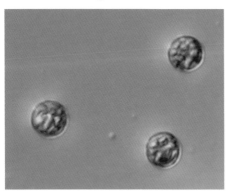

Green algae, a common form of phytoplankton. Each of these unicellular organisms is about 0.01 mm in diameter.
Photo: Alexa Bely, University of Maryland.

Phytoplankton is microscopic and often single-celled. Popularly known as "sea grass" because, like all plants, it utilizes solar energy directly in photosynthesis, phytoplankton forms the basis of the entire marine food chain.

There is a gigantic amount of phytoplankton: 300 billion tons of it are created annually, nearly twice as much vegetable matter as is produced on land. Phytoplankton produces 70% of all atmospheric oxygen. And once these tiny plants die, accumulate on the ocean bottom, decompose, and are transformed through tectonic processes, they are thought to be the main source of the Earth's oil deposits.

Phytoplankton develops in the oceans' sunlit surface zone, converting solar energy and the nutrients dissolved in seawater — carbon dioxide, nitrogen, phosphorous, etc. — into carbohydrates. The process is so efficient that all of these dissolved elements would be entirely used up in just a few days if they were not renewed. Hence, phytoplankton can be abundant only in areas where deep ocean waters rich in nutrients rise to the surface, replenishing the supply. Now the enrichment of deep ocean water is itself due to the continual rain of organic debris from above, the drifting down of wastes and dead zooplankton, fish and other marine animals that live near the surface. On reaching the bottom these wastes and dead organisms decompose, liberating their nutritive mineral elements for use by the next generation of living things. But in order to be available to that new generation, the released elements must be carried back up to the surface by ocean currents. For this reason, plankton and all associated marine life is primarily concentrated near the ocean surface in areas called *upwellings* (Q. 44) where currents carrying water from the depths rise to the surface.

Wherever phytoplankton is abundant, chlorophyll in the living plant cells imparts a green color to the water (Q. 29).

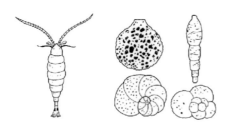

Zooplankton is essentially composed of single-celled organisms such as foraminifers, small crustaceans such as copepods, the eggs and larvae of various marine animals, and fish fry.

Examples of zooplankton: at left, a copepod, at right, foraminifers.

Most forms of zooplankton feed exclusively on phytoplankton. Some feed on other living forms of zooplankton and on their waste.

When shelled organisms die, they sink to the bottom where their mineralized parts form great accumulations of mud. Under pressure, this mud is compacted and becomes fossilized. Such accumulations give rise to certain types of rocks such as limestone, which is composed of tiny foraminifer shells rich in calcium.

39. What kind of seaweed floats in the Sargasso Sea?

When Christopher Columbus found his vessels surrounded by great masses of floating seaweed, he was sure that his voyage was nearly over, so obvious did it seem that such an abundance of marine plant life could only be a sign of land nearby. His crew took a darker view of the situation, fearing that the floating masses would block their further progress and keep them trapped. But the fears of the crew proved to be quite as groundless as the optimistic hopes of their captain.

The sargassum seaweed (from the Portuguese *sargaco*, bunch of grapes) is a type of brown algae that floats in mid ocean, far from land. It never forms masses dense enough to stop a boat.

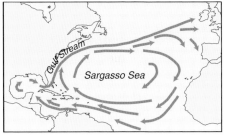

The Sargasso Sea is the name given to the area of the Atlantic Ocean located between the Gulf Stream and the equatorial current. Its remarkably clear blue waters contain a certain amount of plankton but not enough to attract many fish, so its depths are something of a biological desert.

The sargassum, although no more than a thin layer floating on the ocean's surface, provides an oasis for small marine life, nourishing and protecting a population of tiny crabs, shrimp and octopi that cling to it. The creatures are entirely dependent on the seaweed for survival; if they were to lose their grip, they would sink to the bottom 10,000 feet below and perish from the great pressure to which they are not adapted.

A myriad of small fish and crabs find refuge in this seaweed. Photo: David Doubilet, courtesy of Time Life Books [84].

The Sargasso Sea also provides an astonishing gathering place for eels. Attracted by as yet unexplained forces, the eels of North America, Europe, and countries bordering the Mediterranean collect there to reproduce and then to die. And from there the eel larvae of the new generation begin their long journeys back to the continental waters of their respective parents.

40. If fresh water is unavailable, can a person survive by drinking sea water?

Water is necessary for life. An average human adult needs to consume at least one liter per day, either as a liquid or as a constituent of food. A person can survive for several weeks without eating but no more than 6 days without water in the best of cases, and will perish within 2 to 3 days in hot weather.

The water we drink must have a low salt content, however, for strangely enough, a person who drinks too much seawater will die of dehydration. This is because humans and other terrestrial animals have evolved to live by drinking fresh water.

Life began with the development of self-replicating, somewhat unstable molecules. To maintain themselves, these molecules had to acquire some sort of protection against chemicals that could destroy their structures. The cell wall plays this defensive role, isolating the molecules of life from the exterior environment. It is a semi-permeable membrane, one that allows only certain molecules (the ones necessary to maintain life) to pass through, but which blocks the entry of others. Thus oxygen, which is necessary to life, can pass into the cell through the extremely small pores of this membrane, and molecules of carbon dioxide, one of the cell's waste products, can be evacuated through the membrane. Pure water can also cross this barrier, but bigger molecules cannot, nor can salts that are "dissolved" in water, i.e. those adhering to several water molecules, as their effective size is too large (Q. 22).

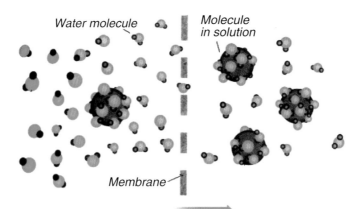

Water molecule

Molecule in solution

Membrane

Water transfer by osmosis

Principle of osmosis. A membrane with very small pores allows small molecules (such as water) to pass through, but blocks dissolved bodies with a larger effective size (such as water plus salt agglutinations). Whenever possible, molecules will cross the membrane until the concentration of the various dissolved substances is the same on both sides. Courtesy of Pearson Education Inc.

If two solutions with identical concentrations are separated by a semi-permeable membrane, water molecules do not tend to cross the membrane, but if the solutions have different concentrations, water molecules will cross over spontaneously, until the concentrations are equalized. This process, called "osmosis," occurs because water molecules are polar (Q. 22), meaning that they have an attraction for ions, such as salt ions, or for other molecules that are themselves polar, such as those of sugar or soap (Q. 203). Water molecules thus form clumps with such bodies, resulting in a shortage of *free* water molecules in the more concentrated solution. Since water molecules are naturally in a state of permanent agitation, they tend to cross the membrane preferentially in the direction of the more concentrated solution so as to equalize the concentration of free water molecules on both sides of the membrane.

When a person drinks seawater, liquids outside the cells become saltier than the liquids inside them. The water molecules inside the cells therefore tend to migrate outward, so as to make the exterior liquids less salty, and the cells become dehydrated.

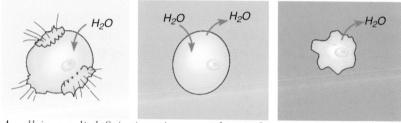

A cell in a salt-deficient environment bursts from osmotic pressure (illustration at left) whereas a cell in an excessively salty environment dehydrates (illustration at right). Normally (center), the rate of water exchange from inside to outside and vice versa is in equilibrium. Courtesy of Pearson Education Inc.

This is what happens to freshwater fish when they swim into salt water: they die of dehydration. And saltwater fish face the same problem, too: their intracellular liquid is less salty than seawater,[1] and they are always fighting dehydration. They drink a lot of seawater to compensate, and eliminate the salt by means of specialized cells. They also avoid losing excessive water by urinating very little.

41. How do fish get the oxygen they need?

Living organisms need oxygen, and water is composed of oxygen and hydrogen, but the bond between these two chemical elements is

---

[1]Remember the accounts of shipwrecked sailors who managed to survive without fresh water by drinking the juice pressed from raw fish.

extremely strong in water. Although it can be broken by electrolysis, this bond cannot be severed by any biological process. Fish can only survive because there is a certain amount of free, elemental oxygen molecules *dissolved* in water. This dissolved oxygen comes from one of two sources: it has either diffused in from the atmosphere or was produced by photosynthesis in marine plants.

Air molecules are in constant motion, bumping into each other and rebounding like so many billiard balls. Any one of these approaching the water's surface will penetrate with ease (remember that matter, whether solid or liquid, is essentially empty space). This mixing process is called diffusion.

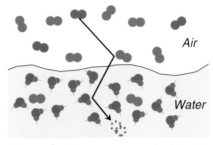

Atmospheric oxygen penetrates water through repeated shocks.

Diffusion is the same process that lets us smell the hamburger cooking on our next-door neighbor's barbecue. If it is windy, of course, that helps waft those aromatic molecules towards us. But even without wind, the smell would eventually reach our noses, dispersed through the air by diffusion.

The second source of dissolved oxygen is photosynthesis carried out by the phytoplankton. Chlorophyll, the "green" molecule that starts the process, absorbs solar energy and passes it along to another molecule, which passes it in turn down through a "chain" of steps, just as a chain of people can pass a bucket of water along from one to another to put out a fire. The details of the reaction are complex, but can be resumed by[2]:

solar energy + carbon dioxide + water $\Rightarrow$ glucose + oxygen.

Note that, even if a lot of photosynthesis is going on, there is a limit to the amount of dissolved oxygen that water can hold. This is called saturation, and it corresponds to approximately 10 milliliters of dissolved gaseous oxygen per liter of water. Beyond that limit, the excess oxygen produced by plankton and seaweed dissipates into the air.

Since photosynthesis requires solar radiation, the process can only take place near the ocean's surface — mainly within the top 30 to 60

---

[2]Or by the formula :

$$\text{photons} + 6CO_2 + 6H_2O \Rightarrow C_6H_{12}O_6 + 6O_2$$

feet — and the diffusion of air in water is also limited to the upper few feet of water. Wave action can drive dissolved oxygen somewhat deeper than that, of course, but, surprisingly, even water from the greatest ocean depths also contains some, the result of the mixing process produced by the great deepwater ocean currents.

Since the solubility of air in water increases as temperature decreases, cold water is rich in dissolved oxygen. The cold waters at the Earth's two poles, being also denser than the surrounding waters, plunge deep (Q. 35), carrying down with them their load of dissolved oxygen. If it were not for this process, life in the ocean depths would be impossible.[3]

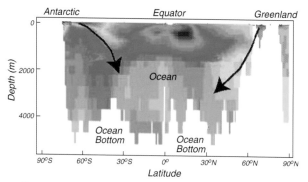

Concentration of oxygen in the Atlantic. Red represents a concentration of 8 ml/l, green, 5 ml/l and blue, 2 ml/l.

There is on average no more than 7 ml/l of oxygen in water. That is only one thirtieth as much as is found in a liter of air (0.7% in water as compared to 21% in air). In order to make use of this dilute concentration, fish have evolved a pair of highly efficient organs, gills, to extract the oxygen and transfer it into their blood.

Gills are composed of thousands of filaments filled with capillary vessels and covered by thin membranes, the total surface area of which is enormous (10 to 60 times the exterior surface area of the fish).[4] Seawater circulates along these filaments and oxygen diffuses by osmosis (Q. 40) into the blood through the thin capillary walls (3 microns thick), going from the water, where it is more concentrated, into the blood, where it is less so. At the same time carbon dioxide,

---

[3]Except near "black smokers," the deepwater hydrothermal springs where certain bacteria and other strange life forms survive without oxygen. It is even possible that life originated in this strange environment, back when the Earth's atmosphere did not yet contain free oxygen (atmospheric oxygen is the product of photosynthesis by plants and certain bacteria, a process which began about 2.5 billion years ago, fully a billion years after anaerobic life had established itself).

[4]We, too, have a large area of contact between our capillary blood vessels and the air: the internal surface of our lungs is approximately 70 m^2 (750 square feet).

a waste product of oxygen consumption, is evacuated in the opposite direction, from the blood into the water.

Gills

Fish gills absorb dissolved oxygen from water by osmosis.

For the system to work adequately, water has to circulate more or less without interruption through the gills. For fish with primitive gill systems, such as sharks, this means they generally have to keep moving. Fish with more evolved systems, however, can "pump" the water, taking it in by mouth and forcing it out through their gills. In this way they can extract up to 80% of the dissolved oxygen.

42. How can fish hover motionless in mid-water without sinking?

Even though heavier than air, an airplane stays aloft because its speed generates a lift on its wings. Sharks make use of this same effect. Although heavier than water, they use their streamlined bodies and pectoral fins to create a lifting force. But a shark that stops swimming will inevitably sink. This also holds true for the sharks' cousins, the skates and rays.

Most other fish have no need to keep swimming because they have a "swim bladder," a sack-like structure filled with oxygen and nitrogen that is connected to the circulatory system. A swim bladder acts like a balloon: if the amount of gas in the

Swimbladder

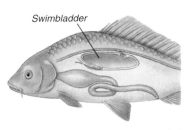

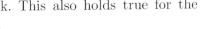

bladder decreases, the fish sinks; if the amount increases, the fish rises towards the surface. Like the captain of an airship or a submarine (Q. 126), a fish adjusts to neutral buoyancy by controlling the amount of gas in its swim bladder. It can then float suspended in the water without moving. Bottom fish, which have no use for swim bladders, have them when young but lose them by the time they are fully grown.

43. Why do fish congregate around sunken wrecks and buoys?

Corals and sponges have to attach themselves to a solid base; they cannot colonize sand. If a bottom is sand or mud, a sunken wreck or a buoy offers such animals the equivalent of a rock to attach to. And once they have established themselves, their presence attracts small fish and other marine animals that feed on them. The food chain is thus established.

Also, just as our own human ancestors found protection in caves, sunken wrecks in shallow water can provide small fish with protection from their predators.

44. What are *upwellings* and why are they so rich in marine life?

An upwelling occurs when water from the ocean depths is carried up to the surface by deep ocean currents. These rising currents bring with them the nutrients that accumulate at the bottom of the sea, and this fosters a teeming growth of plankton in the upper parts of the water column where light can penetrate. A chain reaction is initiated: the proliferation of plankton nourishes a large population of small fish, the small fish attract medium-sized predators, and that brings in the large predators. The world's richest fishing grounds are thus located in zones of upwelling (Q. 113).

Upwellings are mainly found off the west coasts of the continents where the prevailing winds, produced by the anticyclonic (high pressure) zones, blow parallel to the coasts. The wind pulls the surface layers of the water along with it, forcing deeper water to rise up to replace it.

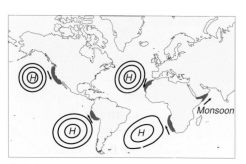 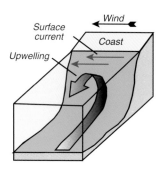

At left, the main areas of upwellings (in dark blue). Upwellings are caused by prevailing winds dragging on coastal waters, which, in turn, cause deeper water to rise (sketch at right).

45. Why is the flesh of tuna fish red ?

All fish have red blood, but the color of a fish's flesh, its muscles, depends on the density of the vascular network in these tissues. Both red muscle and white muscle exist. White muscle is a poorly vascularized tissue and uses stored glycogen as an energy source. This type of muscle lets a fish perform intense, short-term activity without immediately using up the corresponding amount of oxygen. Red muscles in fish are highly vascularized and extract their energy directly from oxygen carried in the blood.

Most fish are sedentary and almost all of their flesh is white, with just two bands of red muscle running along the backbone. This allows the animal to swim continually, but only rather slowly.

But for fish such as tuna that swim both constantly and rapidly, the red muscles are much more highly developed and almost all of their flesh is red.

46. Most fish are cold-blooded. Tuna is not. Why?

Unlike nearly all other fish, the tuna is a warm-blooded creature. Exceptional in many respects, a tuna can reach a length of 10 feet and weigh over 2000 pounds — and it is all muscle! An extraordinary swimmer that can sprint at 30 knots and cross the Atlantic from west to east in 40 days to reproduce, it can also dive to a depth of 3000 feet without suffering from cold.

As any athlete knows, cold muscles work poorly; the optimal temperature for muscle cells is 30 °C. Warm-bloodedness is what we humans owe our strength and agility to, the ability of our bodies to maintain a high enough temperature to let our muscles work optimally, no matter what the external temperature. This is a characteristic that we share with the other mammals and with birds.

A fish is in a much more difficult situation. Since water conducts heat much better than air does, a warm-blooded marine animal has to expend 10 to 30 times as much energy to maintain its temperature as it would if it lived on land. For this reason the vast majority of fish do not regulate their temperature, but simply have the same internal temperature as their external environment. Tuna, however, maintain an internal temperature that is 10 to 20 °C higher than the surrounding water. This allows them to have powerful swimming muscles and puts their cold-blooded prey at a disadvantage. Their hunting zones are vast, too, as a consequence, extending to high latitudes and deep into cold waters. In order to reduce heat loss, tuna have evolved an effective form of internal insulation: their warm red muscles are sheathed in "cool" white muscle tissue.

47. How many shark attacks occur in the world annually?

There are approximately 50 to 70 shark attacks each year worldwide, about 10 of which are fatal. The attacks generally take place in areas where the water temperature is above 20 °C [26]. The highest incidence of attacks is along the coast of Florida, not because sharks are particularly numerous there, but because there are a greater number of swimmers there than elsewhere. The most dangerous region in the world is Australia.

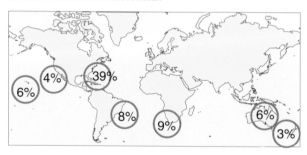

The regions of the world with the most reported shark attacks. Source: Florida Museum of Natural History [21].

As for Europe, the risk is small though not zero: one attack about every 5 years in the Mediterranean, with a fatality approximately once in 10 years. The areas most at risk in that part of the world are the coasts of Italy, Croatia and Greece.

Most attacks occur near a coast or offshore in areas with coral reefs or shallows (51%), Only about 15% of attacks happen in the open seas, but, once again, these statistics do not really indicate the level of risk, only the relative proportions of swimmers and divers in the different areas.

Surfers suffer the most accidents: 49%. Next come swimmers (22%) followed by divers (11%).

To lower one's risk, it is best not to swim or dive alone, not wear brightly colored clothing (sharks are sensitive to contrasts), and not go swimming late in the afternoon or at night (times when sharks are most active). If you should see a shark, try to avoid wild splashing and thrashing around: that sends the creature a signal "frightened prey over here."

48. Do giant squids really exist?

An imagined attack by a mythical animal of the deep.

Giant squids certainly do exist, but they are among the least known of large sea creatures. Until 2005, when Japanese researchers released a series of spectacular still images from a deep water camera showing a giant squid trying to wrench a piece of bait from a hook, no one had ever seen one alive in its natural habitat. A few dead ones had been dredged up in fishermen's nets, and dead or dying ones had occasionally washed up on beaches, usually half eaten by birds or in an advanced state of decomposition. Whalers had also described seeing sperm whales cough up giant tentacles and finding enormous "parrot beaks," the squid's terrible (and indigestible) jaws inside of whale stomachs.

The huge size of the tentacles that washed up greatly stimulated the imaginations of storytellers who told of monstrous squid big enough to attack large ships, feeding the fear of sailors. Jules Verne's Captain Nemo had to weld an axe to fight off a squid attack on his beloved submarine, the NAUTILUS.

Such attacks remain in the realm of myth, however, for the giant squid is a creature that lives at great depths,

Artist's rendering of a giant squid in its natural state.

probably between 1000 and 3000 feet below the surface. Being adapted for great cold and pressure, it cannot rise to the surface without suffering from decompression and the unaccustomed heat. Hence, when one is seen, it is likely to be either already dead or else in very bad shape.

The giant squid (scientific name: *Architeuthis dux*) can reach a length of 60 feet with tentacles outstretched, and can weigh nearly a ton. Like the common squid and certain other mollusks in its family, it has an internal skeleton of sorts, composed of a horny "shell." Its eyes are huge, the size of dinner plates, allowing it to see its prey down deep, where no sunlight penetrates, and only faint glimmers of bioluminescence light the way. It is preyed on in turn by sperm whales (which also eat smaller squid species), which hunt them at great depths using their "sonar" system to track them down [18, 19].

49. Are dolphins fish?

The answer is: some of them are and others are not. The term "dolphin" is applied to two very different kinds of sea creature, one of them a true fish, the other an aquatic mammal.

The true fish — with gills and scales — is now better known by its Polynesian name, mahi-mahi. When listed as such among the entrees on a menu, this avoids confusion with its unrelated namesake, the mammal. A good-sized, active bony fish with tasty flesh, the dolphin fish or mahi-mahi is a native of tropical and temperate seas.

The other dolphin, the mammal, is really a type of small whale. Rather than having gills to extract oxygen from water, it has lungs and breathes air, as all mammals do. And it nourishes its young with milk, too, again as mammals do.

In scientific terminology, dolphins and other whales are categorized as cetaceans. The ancestors of these animals were terrestrial until about 70 million years ago, when they returned to live in the sea.

Since cetaceans have lungs, not gills, they have to come up to the surface periodically to take air in through their nostrils, or blowholes, which are located on top of their heads. The blowhole is directly connected to a whale's lungs but not to its mouth cavity. So unlike ourselves, these animals cannot breathe through the mouth.

All whales are good divers and can hold their breath for long periods of time — in general, the larger the whale, the longer it can go between breaths. Small whales such as dolphins can dive for up to 20 minutes, while the biggest species can remain submerged for over an hour and a half. When diving, a whale closes its blowhole to prevent water from flooding its lungs. On returning to the surface, big whales powerfully expel the air from their lungs, sending out a spray of droplets and water vapor that is visible several miles away, whence the whalers' classic, "Thar she blows!"

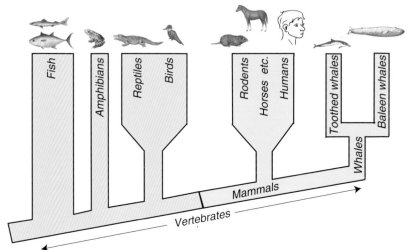

The tree of life for vertebrates. Fish, the first vertebrates, appear 400 million years ago. Amphibians are the first vertebrates to emerge from water to live on land, 350 million years ago, and mammals appear 200 million years ago. Cetaceans, or whales, are land animals that returned to live in the sea about 70 million years ago.

Another major difference between fish and cetaceans is their manner of swimming. Fish, which evolved in the sea, have skeletons that are designed to swim by undulating from side to side. A whale's skeleton was really designed for a land animal walking on all fours,[5] and, like all mammals, whale backbones are built more for flexing up and down than from side to side (think how easily we can bend over forward to pick up something on the ground, for instance, but how hard it is to bend sideways). So whales swim forward by beating their tails up and down, just like a human diver wearing fins (Q. 52).

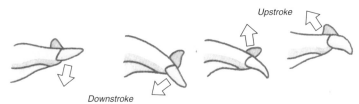

Whales swim by beating their tails up and down (whereas fish undulate from side to side.)

---

[5]In whales, the body has become elongated and fins have replaced the legs. The bones of their former forelegs can still be found inside the pectoral fins, but the hind legs have completely disappeared, and the tailfin that has developed is only cartilage.

50. Blue whales, narwhals, humpbacks, killer whales, dolphins...what makes a whale a whale and how many kinds are there?

Whales are mammals, of course, not fish, which means, among other things, that they nourish their young with milk as we humans do and that their brains are much more highly developed than a fish's. The 80 or so different species of whales (cetaceans) are divided into two major groups: toothed whales (Odontoceti) and baleen whales (Mysticeti).[6]

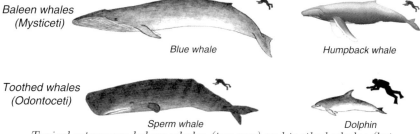

Baleen whales
(Mysticeti)

Blue whale   Humpback whale

Toothed whales
(Odontoceti)

Sperm whale   Dolphin

Typical cetaceans: baleen whales (top row) and toothed whales (bottom row). The divers give the scale of the whales compared to humans.

Baleen whales have no teeth, but their jaws are armed with long horny fringes called "baleen" (also called "whalebone," and once used to stiffen ladies' corsets). Through these they filter great mouthfuls of seawater to strain out their food, essentially krill, which is a planktonic crustacean similar to a small shrimp. These whales breathe through their two nostrils, their double blowholes.

Pygmy whales are the smallest baleen whale, with adults reaching a length of about 15 feet. The largest species is the blue whale, up to 100 feet in length. How strange to think that this gigantic creature, not only the largest animal on Earth today but the largest that has ever lived (yes, even including dinosaurs), nourishes its vast bulk on a thin soup of plankton!

Baleen whales swim with their mouths open, then close them and strain the water out through the fringes to retain the plankton.

---

[6]Scientific names for living things are made up of descriptive Latin or Greek terms. "Odont-" is Greek for "having teeth." "Mysti-" comes from the Greek "mystax," mustache, and refers to the fringe of baleen that hangs from the upper jaws of baleen whales. "Cetus," in Latin, means whale.

Toothed whales are active predators, not plankton strainers, and their jaws can contain up to 200 teeth. Killer whales, sperm whales, narwhals, dolphins and porpoises are all toothed whales. They vary in length from about 5 feet for porpoises to over 50 feet for sperm whales.

These whales prey mainly on squid, crustaceans and fish, which they tend to swallow whole unless they are particularly big. Killer whales, or orcas, are the most aggressive members of this group. They hunt in packs and will attack even seals and the young of other whale species. The male narwhal has the distinction of having a single tooth that grows forward in a long, straight spiral out through its upper jaw and lip, reaching a length of about 20 feet. In the Middle Ages, narwhal teeth were seen as proof of the existence of unicorns and were sold and traded for fabulous sums. Scientists have recently determined that this tooth is not a weapon or a badge of sexual fitness, but a sensory organ that transmits to the narwhal's brain all sorts of information about the temperature, pressure, and composition of the water.

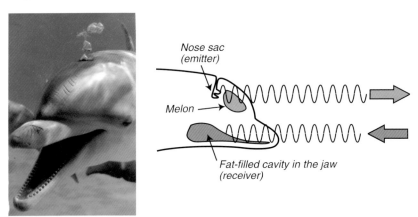

At left, dolphin with blowhole visible. At right, the dolphin's sonar system.   Source: Animal navigation [81].

Whereas the plankton-feeding baleen whales can simply swim along with their great mouths open, swallowing whatever they encounter while breathing through their two nostrils, their double blowholes, the toothed whales, being hunters, need an efficient system for pinpointing and tracking prey even at water-depths where little light penetrates, and they have "sacrificed" one of their blowholes for this purpose. One nostril has thus evolved into a component of an echolocation system that works like sonar, helping them find prey and detect obstacles up to 300 yards away. They emit high-pitched sounds, using the oily *melon* in their foreheads as an acoustic lens, then detect the echoes that bounce back with a specialized structure in their jawbone. They also make complex sounds that probably play a role in social cohesion and group identification.

51. Why do dolphins like to play with bow waves?

The ballet of dolphins playing in the bow wave of a boat is a welcome distraction from the monotony of a long passage.

Most mammals play, especially young ones. This is probably a useful trait. It may help them learn to survive in a group or to hone hunting skills, or even just keep their brains exercised and alert.

But it also looks as though mammals play for the pure fun of it. Just watch a horse gallop friskily across a pasture or a dog enthusiastically playing ball with its master. Scientists like Jacques Cousteau and his crew have often seen dolphins and otters dancing around them in the water, for all the world as if they were looking for admiration and applause [56]. And several recent studies have shown that mammals are not the only animals that play [4]. Certain birds, such as crows, and even turtles, have been seen to do things that are hard to explain with any other word.

Now, talking about play in animals is still something of a taboo in scientific circles. Researchers who do so are likely to be accused of anthropomorphism. But why should animals, some with quite complex brains, lack all sense of pleasure in moving their muscles, jumping, dancing, hiding and finding things, or interacting with each other? Why should this sense of happiness that we call "animal good spirits" not exist most precisely in animals? Why deny them that?

Some students of animal behavior have suggested that, as far as dolphins are concerned, the animals position themselves in a ship's bow wave to save energy. How then do we explain that, out in the middle of the Atlantic, a band of dolphins migrating in one direction rushes over to a boat heading in the *opposite* direction, plays in its bow wave for a good half hour, then resumes its way when the animals get tired of the game?

So it would seem that they position themselves in the bow waves of boats and small ships out of pure pleasure. There, they can ride the wave, carried along like a surfer at 6 knots or more with no need to make the slightest effort. They may have learned the game in the ocean swell, or maybe in the wakes of their own dolphin mothers [68].

Dolphins also like to play with each other. They rub fins or pass a fish or an object around like so many soccer players. They are also probably playing — at least some of the time — when they jump. They have to come to the surface every few minutes to breathe (Q. 49), but they take advantage of the moment to leap high, then plunge back into the water in perfect form, like Olympic divers. Dolphins can make  leaps up to 10 feet high, all the while doing fantastic aerial acrobatics.

52. How do fish swim ?

Fish move forward in the water with a sculling movement, beating their tails from side to side [67]. When the tail moves sideways, the water exerts a reaction force perpen- 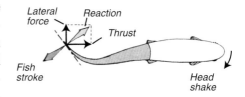 dicular to the tail. The axial component of this force pushes the fish forward, while its transversal component is compensated by a shake of the head.

Generally speaking, since the whole body of the fish except the head works like the blade of a scull, the longer the fish, the faster it can advance, but there are many exceptions. The fastest fish on record is the sailfish, which can speed along at nearly 60 knots over short distances [59].

Water is a particularly easy medium to move through, and since a fish's apparent weight is zero, it has no weight, only friction, to overcome. Also, although the density of water is 800 times that of air, fish with their streamlined bodies come out as the winners. If one measures the energetic "cost" of displacement per pound of body weight transported and per distance covered, swimming turns out to be the most economical means of getting from one place to another.

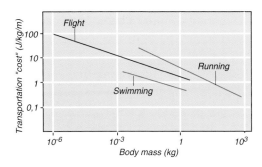

Swimming is the most economical means of locomotion. Source: N.A. Campbell [11].

Certain fish have come up with an improvement on the simple sculling movement, something that scientists discovered only recently.

Von Karman eddies on the windward side of the island of Guadalupe, off the Pacific coast of Mexico. SeaWiFs, NASA satellite image.

When a body moves rapidly through water, eddies form on either side of it. These eddies detach themselves from the body, one after another, first on one side, then on the other. The wake then consists of successive pairs of eddies turning in opposite directions, called "a von Karman vortex street." The forces created by the alternate detachment of such eddies have caused bridges and lighthouses to collapse when they coincided with the resonance periods of the structures. This is a very general phenomenon that even occurs in the atmosphere, as the fine satellite image of cloud formations at left clearly shows.

Certain fish make use of this phenomenon. By moving their tail-fins in synch with the formation of the eddies, they make them work for them. When an eddy forms on the left, they give a single beat to the right with their tailfin. This creates an eddy turning in the opposite direction.

This reverse eddy combines with the natural one to create a propulsive stream, and the fish is effectively pushed along by a rear-mounted motor [77]. The effect allows blue sharks and tunas to attain speeds of almost 30 knots. It is so strong that studies are currently under way to try to reproduce this kind of swimming mechanically, for possible use on submarines.

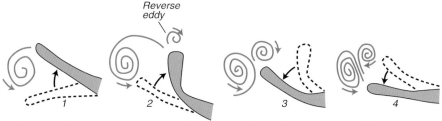

Certain fish create a reverse eddy with a flick of their tailfin (2) which, when combined with the natural eddy (4), considerably increases their speed.

53. Do whales and dolphins ever sleep?

Marine mammals such as seals and manatees can sleep on beaches or rocks, or even stretched out on the bottom of shallow water. But, in the open sea, whales, whether toothed or toothless, would suffocate and sink to the bottom if they fell asleep too soundly. For these creatures have learned the lesson "I must hold my breath" so thoroughly that they have all lost the very reflex of breathing. They can only take in air when they do it consciously. Still, they do manage to sleep. About eight hours a day, even. Dolphins have been observed floating inert on the surface, one eye closed and one eye open. So it would seem that they have developed the ability to let one half of their brains rest, while the other half takes care of regulating breathing. The expression "half asleep" would apply to them quite nicely in that case! But the theory remains to be verified.

54. Why do some fish swim in schools?

Fish commonly swim in groups. Of the approximately 24,000 species of marine and freshwater fish, two-thirds live in schools when young, and 20% continue to do so as adults. The reasons are not yet completely understood, but recent studies have led to several theories:

- Staying in schools helps protect from predators, at least as long as the predator is not too big. A school consisting of thousands of small fish moving in synch looks like one big creature and may intimidate medium-sized predators. The formless mass of the school may also confuse some attackers and make it hard for them to pick out and track a single individual as a potential lunch.

– Schooling behavior may make it easier for all members to find food. There are more eyes on the lookout and, by working as a team, fish in schools might capture bigger prey than any one individual could have caught.

– Schooling fish save energy by swimming in their neighbors' wakes (as bicyclers do during races, or geese flying in V-formation during long migrations).

– Finally, schooling behavior keeps the fish of both sexes together, thus increasing everybody's chance of finding a partner and reproducing.

School of anchovies in the Pacific.
Source: NOAA.

How do schooling fish manage to swim in such perfect synch? Visual cues seem to play a big part: each member of the school notes some physical aspect of a neighboring fish, a fin perhaps, or its tail, or some colored spot on its body. Naturally, this is not possible when visibility is poor. And as a matter of fact, fish schools tend to lose their compact structure at night. Another type of cue used seems to be the detection of the waves made by nearby fish as they swim along. To detect these waves they probably use the sensory cells that run down the length of their bodies on both sides (lateral line system).

55. How do flying fish fly?

Many kinds of fish can jump out of the water, but most fall right back in. Flying fish, however, can remain aloft for almost a minute and "fly" more than 300 feet. That seems almost as hard to believe as the legend of Icarus, doesn't it? But in truth, flying fish do not exactly fly; what they do might better be termed "assisted takeoff gliding."

Their technique is to swim swiftly to the surface, then fan out their "wings" (unusually well developed pectoral and pelvic fins) as they leave the water. Since the speed they can reach by swimming is not enough to glide for any length of time, they accelerate once out in the air by feverishly beating the tips of their tails back and forth on the surface of the water with a sculling motion.

 They glide a few inches above the waves at speeds of about 5 to 8 knots, staying aloft for perhaps 50 or 60 feet, but they can extend that distance by sculling again with their tail. Flying helps them escape from their predators, mainly large surface-living fish and dolphins.

56. Why are there no marine animals larger than the largest whales?

The blue whale is an enormous animal. At 100 feet in length and a mass of 200 tons, the equivalent of 50 elephants, it is by far the largest animal that ever lived on Earth, twice the size of the biggest dinosaurs. Its tongue alone weighs as much as an elephant, its heart as much as an automobile. Some of its blood vessels are so big that a man could swim inside them.

So perhaps it is odd to think that there might be an even larger animal. But after all, why not? The ocean is a particularly favorable physical environment. On land, the skeleton has to support the whole weight of an animal. At sea, on the contrary, marine animals can just float; the skeleton has no need to overcome the force of gravity. Its function is simply to hold the organs together and permit a certain amount of movement for swimming.

Now, if floating brings such an adaptive advantage, why did nature stop with the whale? Why did it fail to produce an even larger animal?

Food supply comes to mind as a possible limiting factor. Let us make a few calculations.

When all is said and done, every bit of food on the land and in the sea is formed from minerals and solar energy. In the ocean, phytoplankton (Q. 38) transforms the minerals in the water into carbohydrates, proteins and fats, which in turn nourish the microscopic animals of the zooplankton. The zooplankton itself serves as food for small fish and crustaceans, and these in their turn serve as food for larger fish, etc. But life is not a perfect machine with 100% efficiency.

Phytoplankton only transforms about 1% of the solar energy it receives into chemical energy, and at each successive step in the food chain the efficiency is only about 10%. This means that a predator can only use 10% of what it eats to grow bigger. All the rest goes to maintain its vital functions and is lost in heat and excrement.

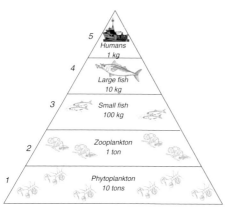

The food chain. It takes ten tons of phytoplankton to produce 1 kg (about 2 lbs) of fish for us to eat.

Now the sperm whale weighs up to 45 tons and needs to consume 1 to 2% of its body weight (i.e. over 1000 pounds) in food every day. It eats large fish and marine mammals, so for this animal the food chain is a long one — about four steps up the pyramid. With a 10% efficiency rate at each step, every calorie the sperm whale absorbs thus requires the equivalent of 10,000 calories of phytoplankton. At that rate, the total surface of the oceans can just barely maintain a population of such whales with enough individuals in it to ensure the survival of the species. So the sperm whale seems to have attained the size limit for what the surface of our oceans can provide an active carnivore. A carnivore the next size up would require 10 times as much food, which means we would need a bigger planet.

This argument does not hold true for baleen whales, however. They are not at the top of the pyramid, but down at level 2, since they feed directly on zooplankton. The blue whale consumes 3 to 4% of its mass per day, or about 2.5 tons. But zooplankton is not in short supply and simple calculations suggest that the amount of plankton on Earth would be sufficient to feed a population of 30,000 "super blue whales," each 10 times as big as the current model. So why don't they exist? The answer seems to be that the lack of any such giant on Earth is likely due to reproductive constraints. Since these whales migrate every year, feeding in the plankton-rich Arctic waters and giving birth in the safety of the tropics, an Earth-year is probably too short to allow for the gestation and lactation period of the corresponding super-giant baby [71, 48].

57. Is coral an animal or a plant ?

A living coral reef looks a good deal like a garden in bloom, with its daubs of bright colors, fronds bowing and waving, and swarms of small creatures flitting about. It is easy to imagine that the coral branches and clumps are underwater varieties of terrestrial vegetation, saltwater cousins of bushes and trees. But they are nothing of the sort. Coral is a form of animal life, and a carnivorous one, to boot! The coral kingdom is as cruel as any jungle: when one coral colony grows too close to another, the two colonies poison and devour each other. But the extraordinary little animals that coral consists of are very simple living forms, way down near the base of the tree of life.

In the whole gamut of living things, where do plants end and animals start, anyway?

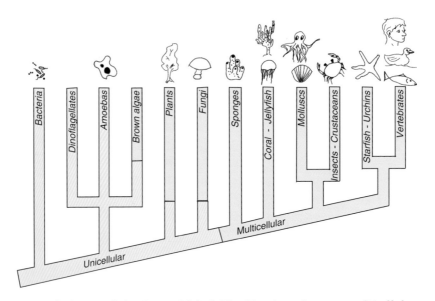

Simplified view of the "tree of life." The blue branches are multicellular. Coral animals are on the same branch as jellyfish.

The main touchstone for this distinction is the way an organism feeds itself. Plants produce their own food using solar energy and basic minerals (by photosynthesis). Fungi represent the next step. They have to absorb organic nutrients created by other plants or animals. They obtain this nourishment directly through their cell walls, by osmosis. So they are generally parasites (living on other living forms) or saprophytes (living on dead and decaying things). Animals come last in line. They "ingest" (eat) their food. Most animals have a mouth and a digestive cavity where the food they take in is broken down into simpler elements.[7] Most animals also have muscles, a nervous system, and can move about.

Sponges are the simplest multicelled animals. They feed on microorganisms by filtering the water. They have neither muscles nor a nervous system, but they can move around. Corals are also very simple but do have muscles and nerves. And like their cousins the sea anemones and jellyfish, they are active predators, capturing small fish and shrimps with their tentacles that are armed with stinging cells.

Thus, corals are animals that look like plants. But paradoxically enough, they cannot survive by behaving like animals. There is a dearth of small prey in their almost barren blue tropical waters, and they have to resort to photosynthesis to survive!

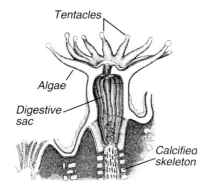

The strategy they have adopted is symbiosis ("living together") with a photosynthetic single-celled alga (a plant, then). This alga, the "zooxanthellum," lodges in the coral's transparent tissues, contentedly photosynthesizing oxygen and glucose which nourish its host. Of all the nourishment corals absorb, 90% comes from their live-in algae guests. As for the benefits to the plant from this symbiotic way of life, it gets a free supply of carbon dioxide (produced as waste by the coral) which plants need for survival, plus a fixed abode for protection.[8]

58. Why are coral reefs found only in warm waters?

Coral reefs cover just a tiny portion (0.1%) of the Earth's surface. They are found exclusively in warm waters, temperatures between 65 and 80°F, and where it is sunny all year long. Most of the reefs are thus located between the tropic of Cancer and the tropic of Capricorn. The presence of a cold current, even in the tropics, will prevent corals from surviving. The west coast of Africa harbors no coral

---

[7]There are some exceptions to these generalizations. Certain plants have lost their ability to photosynthesize and certain animals have lost their digestive cavities.

[8]There are a few species of deep water (from 100 to over 2500 feet) coral without symbiotic algae, where photosynthesis cannot take place because there is so little light. These corals are exclusively carnivorous, like jellyfish and sea anemones.

because of the Benguela current that brings cold water up from Antarctica, and the Humboldt current produces these same conditions along the west coast of South America. On the other hand, reefs can exist in non tropical regions if a warm current is present, as is the case in Bermuda which is warmed by the Gulf Stream.

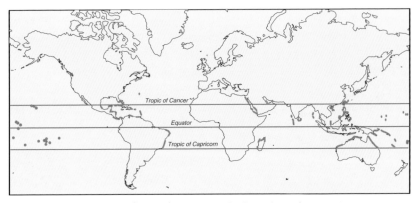

Coral reefs (in red) are mainly found in the tropics.

But warm water and a sunny climate are not the only requirements for corals to thrive. The salinity of the water has to be adequate, too, meaning that river estuaries (the Amazon estuary, for instance) are out. Nor can the water be charged with too many fine particles, or sedimentation would quickly smother the living coral cells. And finally, the water must not be too deep, for at depths greater than about 100 feet, there is too little sunlight for most corals to survive.

Oddly enough, though, the coral reef bases of Pacific atolls and the Maldives islands are 1200 feet below the ocean's surface. The coral at that depth is dead, but what could account for the fact that it was once alive? The answer must be that at one time the living reef was close to the surface, but then was carried deeper as the island that gave rise to the atoll slumped and sank (Q. 59).

A coral reef cannot survive if the sea level rises faster than the reef can grow. Most of the world's coral died about 8000 years ago, when there was a relatively rapid rise in sea level due to the melting of the glaciers.

59. How are atolls formed ?

Atolls are coral reefs out in the open seas. Now, how could a reef have grown up from the ocean floor if coral can only live near the surface? That mystery was solved by Charles Darwin during his famous voyage aboard the BEAGLE in 1832-36. Darwin is mainly known for his theory of evolution, but his contributions to geology were just as extraordinary.

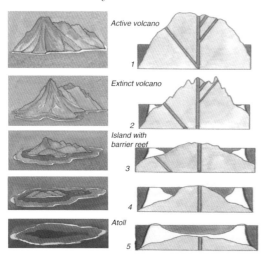

Active volcano 1

Extinct volcano 2

Island with barrier reef 3

4

Atoll 5

He understood that "the islands in the middle of the oceans rise up from the sea floor, born of volcanic activity." Then, once the volcano emerges, coral colonies establish themselves in the shallow water all around it. With time, the island is eroded and slowly sinks, while the coral continues to grow, compensating for its sinking base. This creates the classic "barrier reef." Eventually, the volcanic center of the island completely disappears under water while the coral colonies, which continue to grow, create the atoll with its central lagoon.

This theory was confirmed more than a century later when core samples drilled on the Bikini atoll produced the remains of the original volcanic island at a depth of 6000 feet. The only adjustment Darwin's original theory has had to undergo is that we now understand that the islands do not simply sink under their own weight; it is the ocean floor under them that slumps.

60. Where does the sand on island beaches come from?

On the continents, erosion due to glaciation, running water and wind action produces the sand found on beaches. But mid-ocean islands are mostly volcanic rock and, if erosion produced their sand, the beaches should logically be black. This is sometimes the case (in Hawaii and the Azores, for example), but is far from being the norm. Most tropical beaches have white (or at least light colored) sand.

If we put some of this sand under a magnifying glass we would see small bits of shells and coral. It is the erosion of the islands' coral reefs and the accumulation of shells and skeletons of marine animals that produce these beaches.

Parrot fish grazing on coral. Photo: Deborah W. Campbell.

Wave action, especially during storms, is responsible for most of the coral erosion. But another phenomenon, "biological erosion," is responsible for a significant amount of sand. The term refers to the destruction of coral by animal life, particularly by parrot fish [5, 50]. This fish feeds on algae that grow on the coral, constantly nibbling at and crushing bits of it with its strong beak-like mouth. It swallows the gritty mouthfuls, digests the algae, then evacuates the coral debris in its excrement. A single parrot fish can produce a ton of "coral sand" annually, and a giant member of the family, the buffalo fish, can produce 5 tons a year.

61. Why do barnacles attach themselves to boat hulls?

Sailors hate barnacles[9] like the plague. There they are already, when your boat has only been in the water for a few weeks, glued to the propeller or the hull, wherever the anti-fouling paint has lost its potency. A barnacle-covered hull will see its speed reduced by 30%. Furthermore, the little beasts are hard to scrape off, and it is easy to gash your hand in the process, as a broken barnacle shell can be razor-sharp.

It was once thought that the barnacle was a mollusk because it has a shell, but it is really a crustacean, like its cousin the shrimp. From the time a barnacle hatches until the time it undergoes its fi-

A group of barnacles, one with its legs extended. Source: Sinauer Ass. [9]

nal metamorphosis (it goes through several), a barnacle larva can swim about freely. Then, as the witching hour strikes, so to speak, its last transformation into the adult form takes place. And like Cinderella racing home before the stroke of midnight, it must quickly

---

[9]Barnacles are marine crustaceans, sub-class Cirripedia.

find a solid surface to attach itself to, or it will die. It can use a rock, a piling, a floating tree trunk, a crab shell, a boat hull, even a whale. The barnacle glues itself onto its chosen home head-first, upside down, then slowly constructs its protective shell. And there it remains for the rest of its life, which can go on for another thirty years. The glue it secretes to cement itself down is so strong that scientists have been investigating its possible use in dentistry and bone repair.

Barnacles feed on plankton that they collect by fanning the water with their legs. The Swiss naturalist Louis Agassiz once described them with humor as an animal that stands on its head and eats with its feet. Its reproductive behavior is just as singular. Every barnacle is a hermaphrodite with both male and female sex organs.

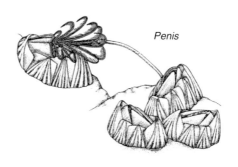

Penis

A barnacle fertilizing its neighbor.
Source: Sinauer Assoc. [9]

But barnacles can neither fertilize themselves nor move around to find partners. With their long, snake-like penises, however, they can grope around, explore, and when they encounter another barnacle who is willing, they deposit their sperm. The barnacle has the distinction of being the animal with the longest penis relative to its size.

62. Are desert islands just a myth?

That depends on what you mean by "desert." Many islands lack a human population, but it would be unlikely to find one without any established land plants or animals. One day or another, a tree trunk harboring insects or worms would wash ashore, a coconut would float in, flying insects would arrive on storm winds, a bird would deposit a seed along with its droppings, and life on land would be established.

Nevertheless, small islands far from continents are very poor in species. As one study done in the Caribbean shows, the number of species an island harbors depends on its size and its distance from the nearest continent. Reducing the surface area of an island by 90% will reduce the number of species on it by 50%.

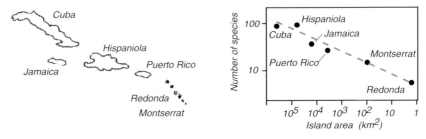

The number of reptile and amphibian species in the Caribbean as a function of island land surfaces [86].

63. For a sailor, sighting a bird is a sign that land is near. What, then, to make of birds spotted out in mid ocean, thousands of miles from the nearest coast?

It is surprising to see birds in mid ocean, several day's worth of flying time away from any land. Some of these are land birds that were carried off by storm winds. Such birds are probably doomed, for if they alight on the water, they cannot take off again. Only birds with webbed feet can do that. But usually the birds one sees are true sea birds, like albatrosses, that spend most of their lives at sea, or else they are migrators flying between continents, like the petrel that crosses the Atlantic. Certain migratory birds (notably the bar-tailed godwit), can remain in flight for up to six days without a rest, covering 5000 miles in a single leg of their journey.

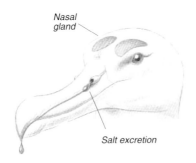

Nasal gland of an albatross.

Sea birds have to drink salt water. To do so, they have evolved special lachrymal glands that help them avoid dehydration (Q. 40). Excess salt is excreted through a tube that opens at the nostrils. From there a channel runs down to the tip of the beak (sea turtles have much the same system, which makes them look as if they are shedding tears when they are out of the water, for instance when the females come to lay their eggs on land). Sea birds have evolved the ability to fly for a long time without stopping to rest. The albatross, with a wing span of over ten feet, can glide for hours in a strong wind without ever having to flap its wings. On the other hand, it has a hard time staying aloft in calm weather and prefers to rest on the water when the winds die down.

64. What are all those slow-swimming turtles doing in the middle of the Atlantic, 1000 miles from land, where they lay their eggs?

Thanks to new techniques of DNA analysis, scientists have just recently begun to understand what those turtles are doing in the middle of the Atlantic, and map the enormous distances they cover in their migrations [7, 43]. They hatch out of their eggs on the east coast of the U.S. between Florida and North Carolina, then immediately launch themselves into the sea to begin a vast migration towards the Azores, Madeira, and even as far as the Mediterranean.

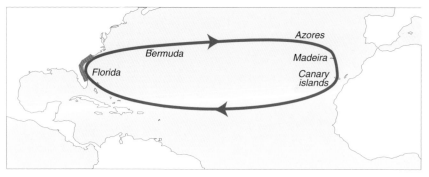

Migration of sea turtles in the Atlantic. The egg-laying zone is in red.

Sea turtles are good swimmers, but they also take advantage of the great current that circles around the anticyclone of the Azores. This current carries them, naturally, to the Azores, where the waters teem with the jellyfish they feed on. The males remain in that region of the sea for the rest of their lives, but not the females, for they have to return to land to lay their eggs — and not to just any land, but to the very same beach on which they themselves hatched out. And so the adult females ride the current back to the coast of America, lay their eggs, then leave again to repeat the cycle. It is likely that similar great turtle migrations take place in the South Atlantic and in the Pacific, too, but this has yet to be confirmed.

What could explain such behavior, swimming such huge distances to regain their native beaches? One hypothesis posits plate tectonics as the reason. Sea turtles are really land animals that became aquatic again about 200 million years ago. At that time, America was very close to Europe and Africa (Q. 7). Mightn't the turtles have developed the drive way back then to lay their eggs on the beaches of America and then continued to do just that, despite the ever-growing distance? At present, that seems to be the most widely accepted explanation.

65. Is there any way to estimate the age of a fish you have caught?

The age of a fish can be estimated by counting the growth rings on its scales,[10] just as the age of a tree can be determined by counting tree rings. Fish grow mostly between spring and fall, and this results in a light band. In winter, they grow very little, which leaves a thin, dark band.

You will need a strong magnifying glass or small microscope (enlargement 10 x to 40 x) to see these rings. Since you probably will not have such items aboard, you will have to wait until you get your fish home. When ready, take a few scales from the middle of the fish's underside (ventral side), soak them in water for a few minutes, rub them between your fingers to remove any traces of adhering skin, and let them dry off.

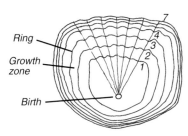

Growth rings on scales. In this example, the fish is seven years old.

Then place the scales on a glass slide, place a cover slip over them, and look at them through the microscope. One pair of light and dark zones on a scale represents a year's growth. Fish have life spans of between a few months for small coral reef fish to fifty years for sturgeons and groupers, and over 100 years for certain deep sea species. The life span of most fish is about 10 to 20 years, with large species generally living longer than small ones.

---

[10] Scientists prefer to count the rings on the otoliths (calcareous concretions in the internal ear), which gives them more precise information, but this procedure has to be done in a laboratory.

The Sky

Skygazing is an activity that anyone can engage in, but sailors have more opportunities to see sunsets and stars than most landlubbers do. Even if the latter wanted to, their view is often obstructed by mountains, trees, or buildings, veiled by smog or degraded by light pollution. At sea, you can see all the way to the horizon, the atmosphere is crisp and clean, and there is all the time in the world...

66. Why is the sky blue?

No need to blush if you don't know the answer to this classic child's question. The phenomenon is far from simple, and it was not until the advent of modern physics that the mechanism was completely understood.[1] To explain it, we need to know two things: (1) what sunlight is composed of, and (2) how it is transmitted in the atmosphere.

Sunlight looks white to us but, as we can see when a light beam passes through a prism, it is actually composed of all the colors, from red through violet and every color in between. Each of these colors is transmitted in small packets of energy called *photons*. The shorter the wavelength of light transmitted the greater the energy of the photon transmitting it: blue photons have more energy than red ones.

When light from the Sun penetrates the atmosphere, most of the photons pass through unobstructed. This is unsurprising, since the atmospheric gasses are thin, containing very little matter and a great deal of empty space. Nevertheless, a few photons do strike atoms in the air molecules they encounter along the way. When this happens, the photon is briefly absorbed by the atom and its excess energy makes the atom vibrate. The atom then releases this absorbed energy in the form of a new photon with the same color (i.e. same energy), but which shoots off in a completely random direction. This light that is re-emitted after a photon collides with an atom is called *scattered light*.

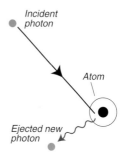

Incident photon

Atom

Ejected new photon

[1]John Tyndall, an Irish physicist, furnished the first explanation of it in 1860, hence the name "Tyndall effect" given to the phenomenon, but the physics of the process was really only understood in the 20th century.

The size of the effect depends on the energy of the photon, in other words, on its color. Red photons, which have relatively little energy, cannot easily penetrate an atom, so most of them just continue on their paths. They are invisible to us because, unless we look directly along their path of arrival from the Sun (in which case we would soon be blinded), they cannot strike our retinas. The more energetic blue photons are more easily absorbed by atoms in the atmosphere, then re-emitted in new, random directions. And some of these re-emitted photons will happen to be scattered directly at us, will strike our retinas, and so be visible to us.

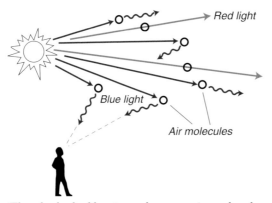

The sky looks blue to us because air molecules primarily scatter the Sun's blue light.

The difference between the number of blue and red photons that are scattered is quite large, varying as the inverse of the wavelength to the fourth power, meaning that blue photons are scattered about six times more than red ones. On the whole, then, of the air molecules that do end up being struck, a very small amount of red light is re-emitted, a tiny bit of orange light, small amounts of yellow and green, a great deal of blue, and an even greater amount of violet. But since our eyes are not very sensitive to violet, our overall impression of the sky is blue.

The events just described can only occur if the particles encountered by the photons are very small compared to the wavelength of light. This is certainly the case for air molecules, which are about 1000 times smaller than the wavelength of visible light. But when photons collide with large particles, they simply "rebound," no matter what their wavelength, so the light that is scattered is composed of all the colors and we see it as white. This effect is easy to see with cigarette smoke. Smoke coming directly out of a cigarette looks bluish because the smoke particles are small compared to the wavelength of light. However, smoke that is inhaled and then breathed out looks white because the particles of smoke become imbued with water vapor in our lungs and are then larger than the wavelength of light. This is why the sky looks milky blue when the atmosphere is charged with fine droplets of water or with dust or pollution, too, and also why the exhaust vapor of a diesel engine is white at first, then becomes colorless after the engine warms up. Those exhaust gasses are a mixture of carbon dioxide and water vapor.[2] At first, the exhaust pipe is cold, the water vapor condenses into droplets before emerging from the pipe, and the exhaust gasses look white. Once the pipe has heated up, all of the water emerges as a gas and the exhaust is nearly invisible.

67. Why is the setting Sun red?

When the Sun is high in the sky, it is *white* — no, not yellow as children's drawings show it. How to prove that? In full daylight the Sun is too bright to look at directly to judge its color, but we can form an image of it with a magnifying glass or photographic lens and project that image onto a sheet of paper: the image is white. The Moon, which simply reflects light from the Sun back to us, looks white, too. The same is true of clouds. Obviously, the color of sunlight for us *should* be white, the neutral color, because our eye was designed to work in sunlight. If our Sun was another type of star, a cooler one, for example, and so redder, our eye would have adapted to this other, redder, light and we would still see it as white.

When the Sun sinks low on the horizon, on the other hand, we can look at it almost directly, but its light has to travel through a thicker slice of atmosphere to reach our eye. As a result, almost all of the blue is scattered, whereas the red and yellow pass through practically intact to strike our retina (Q. 66).

When the Sun has dropped to within 5° of the horizon (the thickness of three fingers held out at arm's length), it passes through 10 times as much atmosphere as it does when directly overhead, and we can look at it without being dazzled: a great deal of its blue light has been scattered and it appears yellow. When it reaches the horizon, the thickness of the atmosphere is 100 times that at the zenith: the blue and green light are strongly scattered and the Sun appears red.

The sky itself becomes red, especially if the atmosphere contains fine particles of dust or water droplets in suspension. These particles reflect the Sun's light and the sky is set ablaze.

---

[2]Diesel oil is a hydrocarbon that, in combination with the oxygen of the air and in the presence of the intense heat of the explosion caused by compression, forms carbon dioxide gas (CO_2) and water (H_2O).

68. Why are sunsets usually more colorful than sunrises?

Throughout the night, water vapor in the air tends to condense because the temperature goes down. By dawn, the air has become clearer, containing less water vapor, and the scattering of the Sun's rays is thus reduced: the sunrise is not very colorful.

In the evening, the opposite is true. Seawater has been evaporating all day due to solar radiation and increased temperatures. By evening the air has become thick with water vapor, which increases the scattering of blue light rays and reinforces the red color of the setting Sun.

Sunsets over land are often more colorful than at sea. During the day the Earth warms up, which strongly agitates the atmosphere and creates updrafts that swirl pollen and dust grains upwards, increasing the number of particles in suspension.

69. What causes the bright rays of light, like searchlights, that stream out from the setting Sun?

Who could fail to admire those magnificent beams of light that sometimes emanate from the Sun near the horizon? Called "crepuscular rays," they appear when light scattering through the atmosphere is blocked in certain directions by clouds or mountains located in front of the setting Sun. The rays themselves correspond to regions of the sky that are illuminated by the Sun, and which stand out brightly against the shadows cast by the clouds or mountains.

Crepuscular rays.   Photo: Holle, University of Illinois.

These rays are particularly noticeable when the air contains water droplets or dust particles that strongly scatter the Sun's light. The rays are actually parallel since they come from the Sun which, for all practical purposes, is at infinity for us. But they appear to diverge from the Sun because of an effect of perspective, just as railroad tracks seem to converge as they approach the horizon.

70. If water and water vapor are colorless, why are clouds white ?

Clouds are made of water droplets suspended in the air, which are large enough to reflect light without changing its color (Q. 66). Since sunlight is white, clouds look white to us during the day. In the evening, however, the Sun reddens (Q. 67) and the clouds take on a reddish color. Clouds can also take on slight colorations in certain other circumstances. It happens when the light that is scattered by a cloud has first been reflected off the land or sea, rather than coming directly from the Sun. In such cases, the cloud can be tinged with the color of the land or water. Eskimos make use of this effect to find areas of open water in ice floes: clouds floating over ice-free water have a slight greenish tinge. The ancient Polynesians navigating on the open seas also used this method to find islands with such low profiles that they cannot be seen directly: over an island with a turquoise lagoon, the clouds are also tinged bluish-green; and above a coral sand island, clouds appear slightly brighter than those over open sea [42].

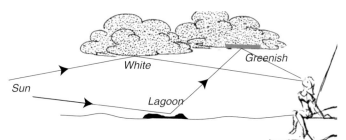

Clouds appear white because they reflect the Sun's rays, but they can sometimes be tinged with the color of the land or sea beneath them.

71. Why are some clouds dark?

Thick clouds such as those charged with rain reflect 75 to 95% of the sunlight that strikes them, so they look white when we see them from the sunlit side. But if we see them backlit, the light from the Sun cannot penetrate and they look dark. The greater their water content, the more light they absorb and the blacker they look.

72. Why do airplanes leave white trails in the sky?

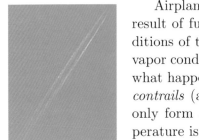

Airplane engines produce water vapor as a result of fuel combustion. Under the right conditions of temperature and pressure, this water vapor condenses into small droplets, exactly like what happens inside clouds. These trails, called *contrails* (a contraction of *condensation trails*), only form at cruising altitude, where the temperature is about -60 °F [6].

Contrails can be used to find one's bearings at sea or even, if all else fails, for navigating. The tale is told that, back when the GPS had yet to be invented and celestial navigation was still required for long ocean crossings, neophyte sailors with an aversion to the sextant would cruise from Los Angeles to Hawaii simply by following the vapor trails left by the tourist planes.

73. Why is the sky sometimes yellowish near the Canary Islands?

Well out into the Atlantic off the coast of Africa, and sometimes as far as the Caribbean, the sky can appear yellowish and the Sun hazy, even in fine weather. This phenomenon, which occurs fairly often, is due to dust from the Sahara Desert that is picked up by sand storms.

The image here, taken by NASA's SeaWifs satellite, shows a particularly violent sand storm in February 2000. The dust cloud stretched out over the Atlantic more than 1000 miles.

Gigantic sand cloud over the Atlantic.

Most of this dust falls into the ocean but, with the right winds, small amounts of it can appear 5 to 7 days later in the Caribbean and even North America. It has been suggested that bacteria, fungus and minerals brought in along with the Saharan dust could be one of the factors causing the destruction of the Caribbean coral reefs [70].

74. Why is the sky white at the horizon, even far from any source of pollution?

On land, any time the sky is less than perfectly clear we tend to blame pollution. And that really is often the case near a city. But out at sea, far from any continent, the lower part of the sky is still white. And the same holds true when astronauts observe the edge of Earth from space.

Out at sea, far from any source of pollution, the sky is white at the horizon. At right, Earth's horizon seen from the space shuttle — there, too, the lower part of the sky is white (the two bright spots in the sky are Mars and Venus). Source: NASA STS 70, 1995.

This phenomenon is due to the fact that the luminosity of the sky, which is produced by the scattering of sunlight by the atmosphere, is stronger at the horizon because we are looking through a thicker "slice" of atmosphere.

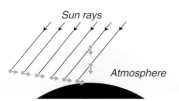

We see more light scattered by the atmosphere if we look in the direction of the horizon than at the zenith.

One might think that, since the sky is blue, it should be even bluer when we look through a thicker layer of air, but that is not what happens. There certainly is a greater probability that a blue photon from the Sun will be intercepted by an air molecule and re-emitted in the direction of our eye, but there is also a greater chance afterwards that it will be re-intercepted and re-emitted in a different direction (Q. 66). The two effects cancel each other out, the sky loses its blue color and appears whitish.

The same thing would happen if our atmosphere were much denser than it is; the sky would be white (and if it were much less dense, it would be black, as it is at very high altitude).

At the horizon, the atmosphere is not just whiter, it also brighter and less transparent. This is what limits visibility when the air is perfectly clean and dry (Q. 174).

75. What is the green flash ?

If we are to believe the old legend that Jules Verne recounts in his novel *The Green Flash*, it is a magic light ray that endows those who are lucky enough to see one with the power to look deeply into their own heart and recognize true love. Well, seeing a green flash may not bring you heightened powers of self knowledge, but its existence is not just a myth, either.

It consists of a brief but unmistakably green flash of light that is produced just as the Sun sinks below the horizon. A relatively rare event, it can only occur if the air is clean and crisp and the horizon is cloudless and well defined. In practice, it is only seen over the ocean. It is caused by atmospheric refraction, and the conditions it requires can be complex.

As anyone who ever navigated by sextant knows, the atmosphere acts as a lens, its density varying with altitude. A beam of light coming in from a star is refracted by this lens and "raised" slightly: this must be taken into account when calcu-

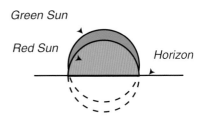

lating a star's true altitude. At the horizon, for example, the Sun is elevated by half a degree. The atmosphere also acts as a prism and so the value of this elevation is related to the color of the light: it is greater for short wavelengths (blue, green) than for long ones (red). The "red Sun" sets first, therefore, and the "green Sun" a split second later — there is hardly any blue in sunlight at the horizon because most of it has been absorbed by the atmosphere. Now, the angular distance between the green Sun and the red one is actually too small for the naked eye to detect, only 20 seconds of arc, whereas we can distinguish nothing under one or two minutes of arc (the limit of visual resolution). So as far as our eye is concerned, the two Suns set at the same time.

The green flash is thus only visible when the effect is amplified. The most common case is a mirage effect rather like what happens on a road in summer (Q. 82) when the sea is warmer than the ambient air. When the red Sun has already set, the tiny bit of the green Sun remaining above the horizon appears to us as if reflected in a mirror, inverted and high enough for us to distinguish it from the horizon. You have to be positioned at least 3 to 5 meters above sea level to see it well. In the middle latitudes, the green flash lasts only a second or two.

A fine green flash.
Photo: Jim Young,
JPL/NASA.

76. Why is the night sky mostly black?

At night, there is no scattering of sunlight to muddle the sky (Q. 66) and it appears almost as crystal clear as if we were out in space. It is black. Now, if you think about it, since the Universe is very, very big, even infinite, and filled with billions of stars and galaxies, if we gaze out in any one direction our line of sight ought to always eventually encounter the surface of a star or a galaxy. It is a little like being in the middle of a good-sized forest: no matter which direction we look, our gaze meets a tree trunk. We have the impression that the forest is limitless and can see no way out. The sky, too, is limitless in appearance, and we know that there are so many stars out there that the night sky ought to be uniformly brilliant. Why isn't it?

This apparently simple question, called "Olbers's paradox" after the German astronomer who formulated it in 1826, had astronomers perplexed for more than a century. Olbers based his reasoning on the hypothesis that matter is uniformly distributed in the Universe. Thus, if we consider all the stars that are equally distant from Earth inside of what we might call a "shell," they send a certain total amount of light energy to Earth or, more scientifically speaking, a given number of photons per second per square meter. If we consider a second shell of stars, larger but equally thick, the volume occupied by the stars in it is larger and the number of stars is therefore greater. So even though each star may appear less bright because it is more distant, we actually receive the same amount of energy from a large, distant shell as from a smaller, closer one. And since the energy received, per second and per square meter is the sum of the contribution of each of the shells, our sky ought to be extraordinarily bright, both night and day.

What Olbers could not know, but modern astronomy has shown, is that the Universe is expanding and therefore has a finite age (about 13 billion years). And Olbers was also making the implicit assumption that the stars never die — but they do. So, if we redo the math, taking the luminosity of the Sun as our average luminosity, taking the average distance we measure between the stars in the vicinity of the Sun as the average distance between stars throughout the Universe, and still supposing that the stars are uniformly distributed but now taking into account the age of the Universe and the age of the stars (which must be less than that of the Universe), we find that there are not nearly enough stars to give us a night sky as brilliant as the Sun.

Then, if we replace the hypothesis of a Universe composed entirely of stars with the modern view of an expanding Universe composed of evolving galaxies, the discrepancy for arriving at a brilliant sky is even greater. The paradox no longer exists.

Isn't it surprising how such a simple question can have ramifications that affect our understanding of the very nature of the Universe?

77. Why does the night sky look darker at sea than when viewed from land?

There is a great deal of "light pollution" on land due to the scattering of our artificial lights by the air and atmospheric aerosols. Lights bathe our cities, and now more and more even the countryside, making the sky gray and masking the stars to the point of invisibility. If you have good night vision and are far from a city, you can see about 10,000 stars with the naked eye. Near a city, you would see only about one hundred. Some city dwellers almost never see the Milky Way!

For the sky to be really black, the nearest small town (e.g. with a population of 2000) has to be more than 10 miles away, and the nearest large city (e.g. with a population of one million) must be over 60 miles away. At sea, that is generally the case. There, the splendors of the night sky which 95% of the inhabitants of industrialized countries are deprived of are all yours.

Moral: if you want to see the sky the way it should be seen, go to sea.

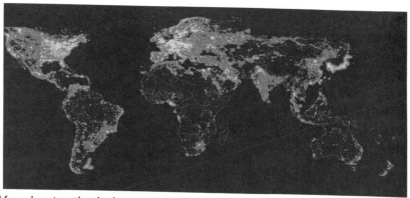

Map showing the darkness of the night sky when seen from Earth. The colors represent the increase of the sky's luminosity as compared to the blackest sky: 10% for dark gray, 30% for blue, 100% for green, and 27 times for red. For those of us who live in the United States or Europe, just about the only places we could go to see a really black sky are the thinly populated, mountainous western states of the U.S. Or else we can go to sea. Based on data from U.S. Air Force meteorological satellites. Source: Cinzano, Falchi, and Elvidge. Reproduced by permission of the Royal Astronomical Society and Blackwell Science.

78. Even though the Sun's rays can still feel hot in the late afternoon, they never produce a suntan. Why not?

The radiation we receive from the Sun is not only composed of visible light, but also of infrared rays that warm us and ultraviolet rays that tan our skin. Ultraviolet rays have a lot of energy,[3] but they are easily absorbed by matter. Just a few millimeters of glass can stop them almost completely. You cannot get a tan through a window.

Air also absorbs ultraviolet rays, and most of those that come from the Sun are thus blocked by the atmosphere. This is certainly lucky for us, since ultraviolet rays are quite deadly due to their high energy — hospitals even use them for sterilizing equipment. In small doses, though, they only destroy a few superficial layers of our skin and cause our capillaries to dilate, producing the inflammation and redness of sunburn. To protect themselves, the deeper skin cells secrete a pigment, melanin (from the Greek *melas*, black), which partly blocks ultraviolet. Tanning is thus the skin's defense mechanism against the aggressive rays of the Sun. The most harmful ultraviolet rays, called "UV-C" rays, have a wavelength of less than

---

[3]The shorter the wavelength, the greater the energy in radiation (Q. 66). Blue light is more energetic than red light, and ultraviolet is even more so.

280 nanometers.[4] They are completely absorbed in the very high atmosphere via the process of ionization and dissociation of nitrogen and oxygen molecules. Rays between 280 and 320 nm, called "UV-B" rays, tan the skin but are still dangerous in large doses. Most of these are absorbed by the ozone layer at an altitude of between 60,000 and 130,000 feet. There is actually surprisingly little ozone in the atmosphere. If one could bring it all together into a single layer, it would only be an eighth of an inch thick. But that is enough to absorb 98% of the UV-B when the Sun is directly overhead. As for the "UV-A" ultraviolet, with a wavelength between 320 and 400nm, i.e., just beyond violet visible light, their energy is lower still and, although they, too, can tan the skin a little, they are fairly inoffensive.

As the Sun goes down, it shines through ever thicker amounts of atmosphere and the thickness increases very rapidly. At 30° from the horizon, its rays pass through twice the thickness of atmosphere that it did when it was overhead; at 10° from the horizon, it goes through 6 times as much and the UV radiation is correspondingly weakened,

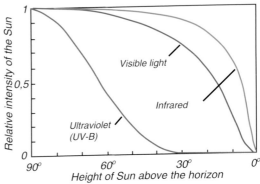

Intensity of the Sun as a function of its altitude. Ultraviolet radiation is negligible when the Sun is below 30° (blue curve), while we can still feel its heat (red curve) [65].

since absorption varies exponentially with the thickness. If, for example, the first meter of a material absorbs 90% of the incident radiation, the second meter will absorb 90% of the remaining 10%, so only 1% will be left. After the third meter, only 0.1% will remain. Thus, the intensity of UV radiation decreases very rapidly with the Sun's altitude.

And so, when the Sun drops from an altitude of 60° to 30° in the course of an afternoon, the intensity of its radiation decreases by a factor of 100. This is why it is impossible to tan late in the afternoon, and why you cannot hope to tan in the winter in temperate latitudes even on the balmiest days. For example in Boston, which is at a latitude of approximately 42°, the Sun rises to an altitude of 65° in June but no higher than 19° at the end of December: an hour's sunbathing in December will not even produce the effect of 2 seconds of sunbathing in June!

---

[4]The wavelength of visible light is between 400 (blue) and 700 (red) nanometers. A nanometer (abbreviation: nm) is one billionth of a meter (10^{-9}).

This also explains why we tend to burn so quickly in the tropics. The Sun rises high in the sky all year long and stays high for hours, so the UV rays are only slightly blocked.

As you can see in the graph, UV-B rays are almost completely absorbed when the Sun is less than 30° from the horizon. So here is the rule to follow if you want to avoid sunburn: *no matter where you are on Earth at whatever time of year, protect yourself from the Sun whenever it is higher than 30° above the horizon.*

If estimating angles is not your forte, just pick up a standard-sized magazine and hold it out at arm's length, as though to read the cover. That gives you an angle of about 30°.

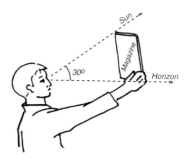

It is good to remember that, although a light cloud cover reduces the brightness of the sky, it does not reduce the ultraviolet radiation much. Also remember that the sea and light-colored sand reflect about 15% of the UV rays, so that on a boat or a beach you can get sunburned even while sitting in the shade.

79. Why does the Moon look so big when it is close to the horizon?

The effect is striking: close to the horizon, the Moon can look one and a half times as large as it does at the zenith. The effect is particularly strong if there are reference objects on the horizon, such as trees or buildings, but it can occur even at sea. You would be disappointed if you took a photo, however: the gigantic Moon would have disappeared. The effect is not real. The atmosphere does not act as a lens, for example, and no physiological structure in our eye is responsible. It really is just an illusion. There is no doubt that the illusion results from an error our brain makes as it interprets our surroundings. The Moon looks quite normal in size if we look at it upside down (by looking at it from between our legs, for example), and there is no illusion in a planetarium, either, even if silhouettes of trees or buildings appear on the artificial horizon.

Several explanations have been suggested for the effect, but none is completely satisfactory. There may well be multiple causes. The most promising theory is linked to our perception of the vault of the sky. It is proposed that we imagine it "flattened" rather than spherical, so that something directly overhead would be interpreted as being closer than if it were at the horizon. That would cause us to subjectively perceive something at the horizon as being larger than the same thing at the zenith, even though it subtends the same optical angle.

Our brain is always making this kind of reevaluation. When we see people walking in the street, the ones in the distance subtend a smaller angle than those we see up close and the images on our retinas are of different sizes. Yet we *perceive* all those pedestrians as being about the same size. Our brain is convinced that those images represent people

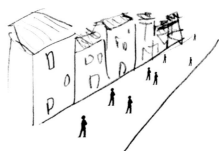

Our brain understands that the small silhouette in the distance is really the same size as the one in the foreground.

and it knows that people have more or less the same dimensions. As a consequence, it corrects our perception. And a lucky thing it does, too, because predators would have made short shrift of the human race long ago if we perceived a tiger in the distance as being no bigger than a tomcat. This same effect is at the origin of the famous Ponzo illusion illustrated below. The two horizontal line segments are equal in length, yet we perceive the top segment as being longer. Our brain interprets the two converging lines as being parallel lines seen in perspective, like railroad tracks. It then draws the conclusion that the upper segment is more distant than the lower one, hence that it must be longer.

Ponzo's illusion

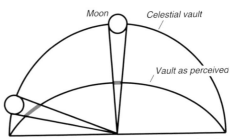

Moon   Celestial vault

Vault as perceived

As in the Ponzo illusion, the Moon appears to be larger at the horizon because we have the impression that the vault of the sky is flattened rather than hemispherical.

The same illusion may be at work in the case of the Moon. The sky is not a vault — it stretches out to infinity. Still, we have the impression that there is a surface above us. And if the Sun, Moon and stars move around on that surface, it can only be a spherical one because the heavenly bodies retain the same dimensions no matter where they are. If it is true that our brains interpret this apparent vault not as a sphere but as a flattened sphere, it could be due to the way we evolved. A survival factor. Food, danger, and protection were to be found more often *around* us than *above* us. It could then be that our brain evolved to favor our perception of things at ground level to the detriment of those high in the sky [76]. Convinced, then, that the horizon is farther away than the zenith, our brain automatically "corrects" our perception of the rising Moon, making it seem larger to us, just as it corrected our perception of the distant, catlike silhouette to warn us that it was really a tiger.

80. When we see the crescent Moon in daytime, its axis does not seem to be directed towards the Sun. Why not?

The bright, sunlit side of the Moon must, naturally, face the Sun, and the axis of its crescent really is directed squarely at the Sun.

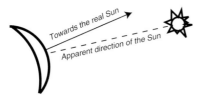

The crescent Moon faces the Sun directly, yet does not appear to do so.

Yet when we see a crescent Moon by day (in the afternoon for a crescent after the new Moon, or in the morning for a crescent preceding the new Moon) and we mentally trace its axis, it does not seem to be directed at the Sun. This is an optical illusion that may be due to our perception of the vault of the sky being flattened rather than spherical (Q. 79). You can make the illusion disappear by taking a long ruler or gaff and lining it up with the bisector of the crescent: the ruler does indeed pass through the center of the Sun [37, 49].

81. Why do the Sun and Moon sometimes have halos?

The halos we see around the Sun or Moon when the sky is hazy are caused by ice crystals in the cirrostratus clouds refracting the light rays.

Cirrostratus clouds are thin formations found at altitudes of 20,000 ft and up, where the reigning temperature is around 0°F. They are composed of minute crystals of ice (ranging from 50 to 100

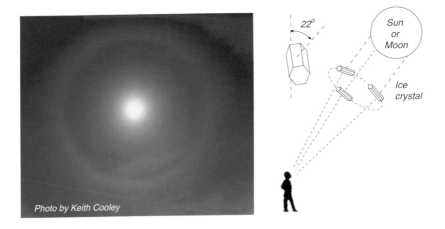

Photo by Keith Cooley

microns), most of which are rod-shaped with hexagonal bases, and they act as prisms. Inside the cloud their orientation is random, so they refract incident light in all directions, but the emergent light is most brilliant in the direction of the minimum angle of deviation of the prism, which is 22°. Thus, the crystals which happen to be oriented in the direction of 22° from the Sun or Moon reinforce their light and create the halo we see. And since ice disperses light just as glass does, the halo is slightly iridescent, reddish on the inside and bluish on the outside.

This halo, which is seen as a detached *ring*, should not be confused with the halo only a few degrees in diameter that one sometimes sees around the Sun or Moon in a cloudy sky, and that is called a "corona." Coronas are caused by the presence of water droplets in clouds. But the effect here is not due to *refraction* as is the case with halos or rainbows, but to *diffraction*: the light coming from the Sun or Moon "bends"

Moon corona.   Photo: Mats Matts- son.

slightly around the water droplets, enlarging the apparent diameter of the celestial body by several degrees.

82. What is the "Flying Dutchman" mirage?

The legend of the flying Dutchman that Richard Wagner transposed so magnificently into operatic form is one of the most famous legends in maritime folklore. It originated in the 15th century, at the very beginning of the era of great navigations, at a time when the difficult passage around the Cape of Good Hope was particularly feared by seamen. According to the legend, Van der Decken, the captain of a Dutch vessel, boasted that he could sail around the cape with the winds against him. For his arrogance he was condemned to fight those winds for all eternity, and his phantom vessel still haunts the waters there, floating in mid air, its sails blood red. Any vessel approaching it is doomed to sink beneath the waves...

This legend probably originated in a perfectly real optical phenomenon, the *superior mirage*.

When light passes from one environment to another, from air to glass, for example, its speed changes, and that causes its direction to change: the light is "bent," refracted. The same phenomenon occurs when light passes through layers of air with different temperatures. Since the index of refraction of air varies with temperature, the light rays are bent, and the object we are looking at seems to be somewhere that it is not.

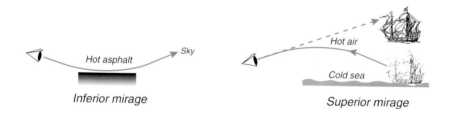

Inferior mirage

Superior mirage

Everyone has seen the mirage that appears on the surface of an asphalt road overheated by the Sun. The air at the level of the roadway is hotter than the air above it and its index of refraction is smaller, so that light rays passing through it are refracted. We think we see water on the road, but it is the sky that we see. This classic mirage is called an "inferior mirage." A superior mirage is produced by this situation in reverse. It occurs on hot spring days when the sea is still cold, making the air at the surface of the sea cooler than the air above it. The difference of temperature in the two air layers bends the light rays is such a way that a ship in the distance appears to be raised above the horizon, seeming to float in air.

When the temperature difference between the two air layers is large, the light rays coming from the bottom of the object are bent more that those from the top, and the object will then appear upside down. Sometimes the ob-

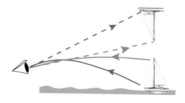

ject itself is visible at the same time as its mirage, since some of the light rays will travel directly towards our eye.

83. What causes the "floating island" mirage?

The floating island mirage is produced when the air is warmer at sea level than above it. The light rays are bent as in the classic inferior mirage (Q. 82), so that the object seems to be near the water as if reflected in a mirror. The upper part of the island thus appears, with its own image compressed and inverted just below. If the island is far enough away, the upper part of it together with its inverted image can appear to float above the horizon on what looks like water but is actually the image of the sky above the island [49].

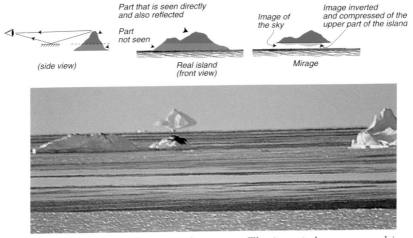

Fine mirage of an iceberg in the Antarctic. The inverted, compressed image of the upper part of the iceberg is clearly visible.   Photo: Guillaume Dargaud.

84. Why are rainbows seldom seen in the middle of the day?

As everyone knows, rainbows form when sunlight lights up raindrops or mist. But the Sun cannot be just anywhere: it has to be *behind* the viewer. A rainbow is actually nothing more than sunlight reflected and refracted by water droplets. The decomposition of light into its components (violet, indigo, blue, etc.) occurs because each drop acts like a little mirror with a prism in front of it. The Sun's rays penetrate the drops and reemerge after undergoing two refractions (hence the prism effect) and one complete reflection.

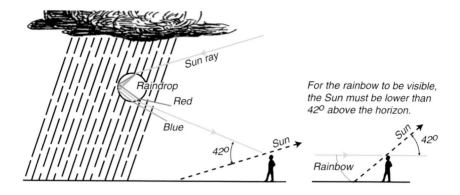

For the rainbow to be visible, the Sun must be lower than 42° above the horizon.

Each ray of a given color makes a specific angle with the incident direction, about 42° on average. Each raindrop lit by rays coming from the same direction sends rays of the same color back at the same angle. The rainbow is thus located on a cone that has our eye at its apex and a half-angle at the summit of 42°. Generally speaking, the rainbow will only be visible if it is seen against the background of the sky, and thus if its highest point is above the horizon. For that to occur, the Sun has to be less than 42° above the horizon. This is always the situation in winter, but at other times of the year the Sun rises much higher. In those cases a rainbow will not be visible in the middle of the day, but only in the earlier part of the morning and later part of the afternoon.

A rainbow is only clearly visible when the water droplets are big enough, with a diameter of about a millimeter. A rainbow cannot form with fog because the droplets in suspension are too small (about 5 hundredths of a millimeter).

85. Why does the Moon always keep the same side facing Earth?

The Moon always presents the same side to us because it spins around on its axis in *exactly* the same amount of time as it takes to revolve around Earth: 27.3 days (the period between two full moons is a little longer, 29.5 days, because Earth has also revolved

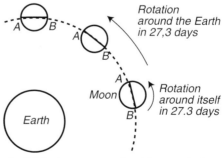

around the Sun during that time). This is not a coincidence; it is a tidal effect. But here it is a tide created on the Moon by the Earth.

The Earth is 80 times more massive than the Moon. It therefore creates much stronger tidal forces on the Moon than the Moon subjects Earth to. There is no water on the Moon, but these tidal forces are strong enough to change the shape of the Moon, causing it to bulge out on the side facing Earth.

In the past, the Moon used to spin faster on its axis and the bulge moved across its surface just as the tide wave does on Earth. This con-

tinual distortion of the surface led to a loss of energy through friction, causing the Moon to slow down progressively. The process stopped when the Moon was spinning on its axis in precisely the same amount of time as it revolves around Earth, and the Moon's bulge became stationary.

Our Moon is not unique in this way. Most satellites of other planets (notably those of Mars, Jupiter and Saturn) also always show their planet the same face. Only satellites that are far enough away from their planet to be free of most tidal effects do not do this. What happened to the Moon will happen to us, too, eventually. Our daily tides are causing the Earth to lose energy through friction between the ocean waves and seashores. The Earth is rotating more and more slowly and the days are getting longer by about 1.8 milliseconds per century. In the age of dinosaurs, a day was only 20 hours long, now it is 24, and in a few billion years, the Earth and Moon will be locked together, each always presenting the same face to the other, and a day on Earth will be as long as a lunar month, about 50 of our present days.

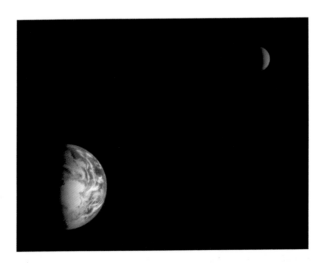

A magnificent view of the couple Earth-Moon taken in 1998 by a NASA space probe traveling out towards the asteroid belt.

86. Without a telescope, is there a way to tell a planet from a star?

Every self-respecting sailor ought to have some familiarity with the sky. Even if celestial navigation has become superfluous in the era of the GPS, we can find pleasure in what was once a necessity and is now only a game, and we can better appreciate the exploits of the great navigators of the past and the debt that the art of navigation owes to astronomy.[5]

Someone who gazes at the night sky only occasionally can find it hard to pick out the planets from among all the myriad of brilliant stars. Four planets are visible to the naked eye: Venus, Mars, Jupiter and Saturn. Venus, being closer to the Sun than we are, can only be seen in the evening and early morning, 3 hours after sunset and 3 hours before sunrise at most. So look for it in the west in the morning and in the east in the evening. Venus is the brightest celestial body in the night sky except for the Moon.

When in the nighttime sky, Mars, Jupiter and Saturn might be seen at any time during the night. Look for them along the path that the Sun travels by day, the ecliptic, and nowhere else. Mars shines with an orange-tinged light and Jupiter is particularly bright, like a brilliant star. Saturn is the least bright of the planets visible to the naked eye.

When in doubt, there is a good way to confirm that you are looking at a planet rather than at a star: stars scintillate, planets do not. Remember "twinkle, twinkle, little star"? The twinkling occurs because our atmosphere is never completely still; the air is always in movement thanks to winds and updrafts. And since the temperature of air also varies, its density — and consequently its index of refraction — varies all the time, too. It is

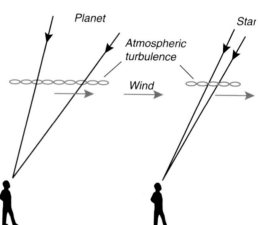

The planets do not twinkle because the angles they subtend are much larger than the size of a turbulence cell in the atmosphere (the angles in this drawing are greatly exaggerated).

---

[5]And vice versa. The first great national observatories in Europe, such as Paris and Greenwich, were established to help improve navigation.

as though a great many little lenses were shifting around above us, blown about by the winds. And so the light we receive from the stars looks briefly, at times, a bit concentrated and at other times a bit dimmer.

Now, the stars are so far away that we can never see their disks; they appear to us only as points of light — even through our most powerful telescopes. But each point of light appears more or less brilliant as turbulence cells in the atmosphere focus its light or cause it to diverge. The stars twinkle.

The planets, on the other hand, are relatively close to Earth and we can make out most of their disks with binoculars or a small telescope. It's harder with the naked eye, but they can also be recognized by the fact that their light does *not* scintillate. The reason for this is that the light from different parts of a planetary disk passes through many different "atmospheric lenses" before recombining in our eye.[6] What we perceive for a planet is its average luminosity, and thus the fluctuations are very attenuated.

---

[6]The angular size of a turbulence cell in the atmosphere is about one second of arc, whereas the apparent diameter of the planets is always larger. Jupiter can have a diameter of up to 45 seconds of arc.

Wind and Weather

87. Why doesn't wind blow directly from high pressure to low pressure areas?

Wind *does* blow from high pressure to low pressure zones; it just doesn't travel along "the shortest distance between two points." And what keeps wind from traveling in a straight line in this case is the Coriolis Effect. Remember that one?

It can be best illustrated by thinking about what happens if you shoot off a cannon from the North Pole. Since the Earth turns while the cannonball is in the air, the projectile will follow a curved path with respect to Earth, and will land, not on the point you aimed at, but to the west of that point (rotate an orange, and mark it with a pen as you turn it to prove this to yourself).

But if you shoot off the cannon ball from any other point on Earth, the results are harder to imagine — the rules governing a world that spins around on its axis are not easy to grasp. Try thinking of what would happen if the world were two-dimensional, like a flat sheet of paper. Imagine that you and a friend are sitting on a revolving merry-go-round. You throw your friend a ball. The ball starts out in his direction, but during its flight the merry-go-round turns and the ball will miss him.

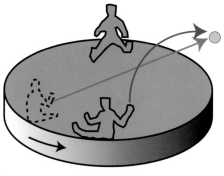

Seen from outside the merry-go-round, the ball's trajectory is a straight line (red trajectory), but seen from inside, it is a curve (blue trajectory). If the merry-go-round is turning counterclockwise, the ball will seem to curve to the right. This is essentially what happens on Earth, but the situation is somewhat more complex because the merry-go-round we ride on is three-dimensional. The following figure illustrates the situation. As before, a ball thrown from a point on Earth continues in a straight line when viewed from space, but its trajectory is curved when viewed from Earth. The deviation is to the right in the Northern Hemisphere and to the left in the Southern Hemisphere. Changing the direction the ball is thrown does not alter this effect, but changing its speed or latitude does make a difference.[1] The effect is greatest at the poles and non-existent at the equator. Since this effect results in a change in a body's trajectory, we treat it as a force mathematically. But, strictly speaking, it is an imaginary force because it has no effect on stationary bodies on Earth, only on moving ones.[2]

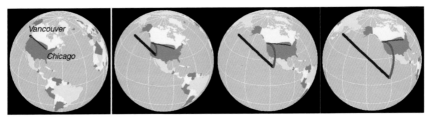

Trajectory of a projectile traveling from Vancouver towards Chicago. In blue, the path seen from space. In red, the same path seen from Earth. Simulation by David MacIntyre, Oregon State University.

The Coriolis force is a weak one, significant only for bodies moving at high speed over long distances. For example, at the latitude of New York a jet plane flying in a straight line would be displaced by about 60 miles to the right after an hour of flying time if the pilot neglected to correct his route.

There is a popular myth that this force controls the direction that water swirls when it drains out of a bathtub or sink, but it is insignificant at such low speeds and short distances. If you repeat the experiment often enough, you will see that the water rotates clockwise just as often as counterclockwise. You might even have the occasion to see the direction reverse itself while the water is draining out. The Coriolis effect can be seen in a tub only if, after filling it, you wait several days for the water to stabilize completely, then let it drain out drop by drop to give the force enough time to develop.

To come back to the original question, large wind systems arise because of differences in pressure between high pressure zones and low pressure zones which can be thousands of miles apart. Over such distances, the Coriolis force has plenty of time to manifest itself, and its influence actually dominates.

---

[1] The Coriolis force is proportional to the mass of the body, its speed, and the sine of the latitude.

[2] The same is true of centrifugal force, which is also an imaginary force.

The air of the high pressure system begins moving towards the low pressure area but is slowed down by friction against the surface of the Earth for one thing and, for another, it is subjected to the Coriolis force which acts perpendicular to its direction. Instead of continuing directly towards the low pressure area, perpendicular to the isobars (zones of equal pressure), it deviates to the right (in the Northern Hemisphere)

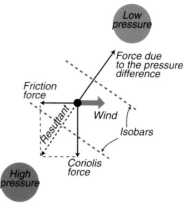

and ends up taking a direction that is practically parallel to the isobars. The angle it adopts with respect to the isobars depends on the wind's speed and the resistance it meets due to friction. Typically, it is 15° over water and 30° over land.

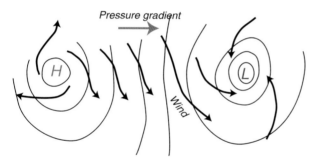

Instead of moving directly from the high pressure to the low pressure area, the wind blows almost parallel to the isobars.

88. Why do winds blow from the east in the tropics but from the west in the middle latitudes?

It's tempting to explain prevailing winds simply by the rotation of the Earth. Our planet turns from west to east, dragging its atmosphere along with it thanks to friction. If the atmosphere lags somewhat behind, that should create an apparent wind coming from the east. This would explain the trade winds that blow roughly from the east. But then, how to explain the west winds in the middle latitudes?

Actually, although the planet's rotation certainly is a factor here, it is not the main one. Temperature differences at the Earth's surface are what put the atmosphere into motion. The surface is warm at the equator, where the Sun's rays arrive almost perpendicularly, and it is cold at the poles, where they arrive more obliquely. The excess

heat in the equatorial areas is transferred towards the upper latitudes
by the two things that are able to circulate on Earth: the air of the
atmosphere and the water of the oceans. In fact, ocean currents and
atmospheric circulation share this work of transferring heat almost
equally: 60% for the atmosphere and 40% for the ocean currents. In
both cases the mechanism is the same as the one that make smoke
go up a chimney: a heated fluid is less dense than a cooler one and
tends to rise, and, inversely, a cool fluid tends to sink.

As you might logically deduce, the overheated air of the equator
rises and is replaced by cold air from the poles, forming what is called
a "convection cell."

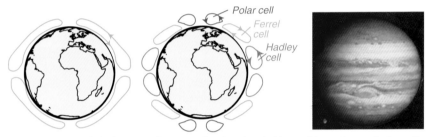

*At left, a simplified view of atmospheric circulation. At center, what really
happens, with 3 convection cells. A similar situation is found in Jupiter's
atmosphere, but with more cells (the light bands correspond to warm parts
of the atmosphere that rise, and the dark bands to cold parts that descend).*

This view is too simple, of course. The small surface area of the
poles could not possibly absorb the huge volume of warm air converging on it from the great perimeter of the equator. The poles are
a bottleneck blocking the warm air's arrival, forcing the atmospheric
circulation to break down into several convection cells. There are actually three of these cells per hemisphere, each one transporting as
much as it can of the excess heat from the equator to higher latitudes. The first one, extending from the equator to 30° N or S, is
called the Hadley cell; the second one, called the Ferrel cell, extends
from 30° to 70° of latitude, and the third one is simply called the
polar cell.

Each cell rotates in a direction opposite to that of its neighbor, a
little like a set of gears in a motor: the rising or sinking movement in
one cell pulls the air of the adjacent one in the same direction. Since
the hot air at the equator must rise and the cold air at the poles
must fall, the directions of rotation of the entire system can only be
as you see them in the figure.

The Earth's rotation on its axis will perturb this movement of air
by the Coriolis effect (Q. 87), at least as seen from Earth. A moving

parcel of air curves to its right in the Northern Hemisphere, and to its left in the Southern Hemisphere. In the Northern Hemisphere, the air in the Hadley cell, which comes from the north when near the surface of the Earth, thus curves right, to the west. This creates the trade winds that blow so regularly from the north-east.[3] In the mid latitudes, the air in the Ferrel cell, which, at the surface, comes from the south, curves right, to the east. This creates the westerlies. The same situation is found in the Southern Hemisphere, this time with the winds deviating to the left: the trade winds blow from the south-east and there are westerlies again in the mid latitudes.

Between the Hadley and Ferrel cells, at about latitude 35°, the air is slowly descending. This is the well known zone of calms (also called the *horse latitudes*). The air's sinking movement compresses it, causing it to warm up (compressing a gas makes its temperature increase) and also dry out. It rarely rains in these latitudes. On the continents, these are the desert areas and, at sea, the subtropical anticyclone zones (high pressure zones) where clouds that may occur are usually too thin to bring rain.

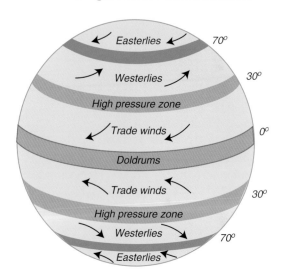

This view of atmospheric circulation is still somewhat simplified, since the presence of the continents modifies it a good bit. The continents block the circulation and, being warmer than the seas in summer and cooler in winter (at least above latitude 35°), they also modify it. Specifically, they cause the band of air between the Hadley and Ferrel cells to break up into a series of high pressure areas, such as the highs of the Pacific and of the Azores.

---

[3]The "trade winds" are not, as one might imagine, winds that were useful in trade in the sense of commerce. Trade is an Old English term meaning even, stable, regular.

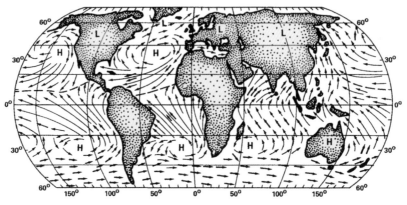

Surface winds over the oceans in August

89. Why are the winds in the "roaring forties" so violent?

The term "roaring forties"[4] refers to the powerful west winds found in the region between the 40th and 55th parallels of the Southern Hemisphere, in the great southern ocean, the only ocean that circles the Earth, connecting the Atlantic, Pacific and Indian Oceans.

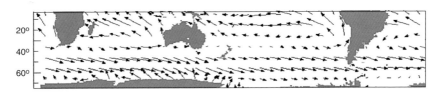

Average direction of winds in the Great Southern Ocean.

These winds have the same origin as the west winds in the Northern Hemisphere (the Ferrel cell), but their average speed is 40% greater. The reason for the extraordinary force of the winds, which exceed 25 knots 50% of the time, is the continual series of strong depressions that rage in this area. These depressions, caused by cold air from the Antarctic encountering warmer air over the ocean, travel from west to east in the great expanse of ocean without being in the least weakened by any land mass.

---

[4]The expression dates from the age of great sailing ships that returned from Australia and New Zealand via Cape Horn. Since the winds increase in violence the further south one goes, the term "roaring" was felt to be inadequate for latitudes beyond 50° and the expression "furious fifties" was coined.

90. How do storms in the mid latitudes differ from hurricanes?

On satellite images, a hurricane (or tropical storm) strongly resembles a mid latitude storm. Both are depression systems that pull their winds into rotation (counterclockwise in the Northern Hemisphere). Their energy in both cases comes from the condensation of water vapor, creating the heavy cloud cover that accompanies them.[5] And their function is also the same: to dissipate and redistribute the excess heat that accumulates in the hottest areas of Earth.

At left, a winter depression off the U.S. coast ; at right, hurricane Hugo *approaching Florida.*

However, since both types of storms are carried along by the general circulation of the atmosphere, they move in opposite directions, hurricanes towards the west and mid latitude storms to the east. And, most importantly, their genesis is different. Hurricanes are "bottom up" events, born of the massive evaporation of tropical seawater, whereas mid latitude storms are "top down" events, initiated by perturbations in the jet stream (Q. 92). Wind distribution is not identical, either. In tropical storms the strongest winds are closest to ground level, while, in the case of mid latitude depressions, they are high up (near the tropopause).

Hurricanes are also very different from tornadoes. A tornado is a strongly rotating column of air attached to the base of a thunderstorm and extending to the ground. Tornadoes form over land (although they can extend over water and become waterspouts) while hurricanes occur only at sea. The diameter of a tornado is only about 300 feet, while hurricanes extend over hundreds of miles. A tornado lasts only about 15 minutes while hurricanes can be active for weeks.

---

[5]Water actually plays a large role in the heat transport. Heat energy from the Sun that served to evaporate the seawater is restored when the water vapor condenses into droplets in the clouds. The air warms up and, as warm air is lighter, its upward movement is accelerated. The surplus of energy acquired at sea level is thus evacuated at high altitudes and towards the poles.

91. If the Coriolis Effect causes winds to veer to the right in the Northern Hemisphere, why do hurricanes turn counterclockwise?

Since the Coriolis force causes winds to deviate to the right in the Northern Hemisphere, it might seem reasonable for hurricanes to rotate clockwise. Actually, they do the opposite. That is also true of winds in mid latitude low pressure areas.

The explanation is that the Coriolis effect is neither the only factor involved nor does it act on stationary objects (i.e., it does not create movement of itself). What does create wind is a pressure difference between two points, what is called a *pressure gradient* (Q. 87). Near the center of a low, the pressure gradient increases and forces the streams of air to deviate and then to begin rotating around it.

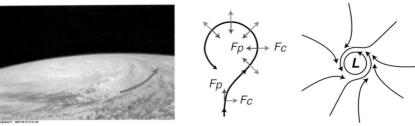

At left, typhoon Winnie approaching the Marianas in the North Pacific in 1997 (Source: NASA, STS-85). The winds rotate counterclockwise. At first, the Coriolis force (F_c) acts perpendicular to the force of the pressure gradient (F_p) and causes the stream to veer right, but as it nears the center of the low the two forces end up in opposition, forcing the stream to take a spiral course. At right, overall view of the effect.

92. How do mid latitude storms form?

Since we normally encounter our weather somewhere down around sea (or ground) level, we tend to think that this is where the air masses and high and low pressure areas meet, sometimes to clash. But it is actually the high altitude winds, the jet streams, that govern our weather at ground level. Discovered in 1944 during the first high altitude flights by American B-29's en route to Japan, these winds are extremely strong and influential.

There are two, more or less clearly defined jet streams in each hemisphere, the polar jet stream at an altitude of about 30,000 feet and the subtropical jet stream at 45,000 feet. No more than thin ribbons of air, they are only a few hundred miles wide and about 3

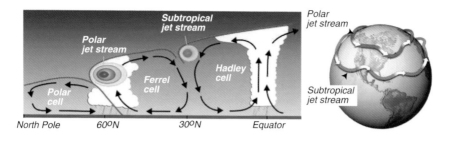

miles thick. The polar jet stream is located at the frontier between the Ferrel cell and the polar cell (Q. 88). The subtropical jet stream, positioned between the Hadley cell and the Ferrel cell, is discontinuous, less strong, and can even disappear entirely in the summer. Under the influence of continental land masses, jet streams follow undulating paths around the Earth rather than simple, circular paths.

These strong, high-altitude winds arise from the great temperature differences between the various layers of air in adjacent convection cells, resulting in large differences in pressure and, hence, in a very rapid displacement of massive amounts of air. Under the influence of the Coriolis effect, these moving air masses turn west to form the powerful jet streams. Since the temperature differences are greater in winter than in summer, the speed of the jet streams is also greater in winter, up to 300 knots for the polar jet stream. The subtropical jet stream is slower because the thermal contrast there is less pronounced. The polar jet stream moves at such high speed that its influence can be felt all the way down at the surface of the Earth. Located as it is at the junction between two convection cells, any change in its path causes the cold air mass of the polar cell and the warmer air mass of the Ferrel cell to move. It is these fluctuations in its trajectory that cause the variations in temperature and the formation of fronts and depressions that we, at the surface of the Earth in the mid latitudes, experience as our weather.

The jet stream does not maintain a constant speed along its path; its speed changes in the meanders, just as the speed of a river current varies in a meandering river bed. The air is compressed in some places and thinned out in others. Where the air "piles up," we say there is a

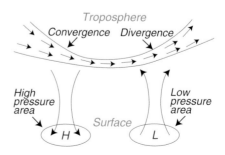

convergence, and where it thins out, a *divergence*. The accumulation of air in convergences tends to make the air sink, whereas divergences create updrafts from the surface. The first case causes a high pressure zone at the surface, and the second, a low pressure area.

Air at the Earth's surface is attracted to an area of low pressure and, thanks to the Coriolis effect, begins rotating as it comes in. This distorts the interface between the warm and cool air masses of the two convection cells, creating the classic warm and cold fronts associated with low pressure systems.

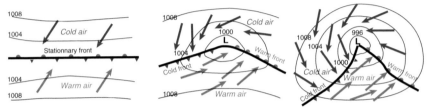

Evolution of a low pressure system. At left, stationary warm and cold air masses. The updraft caused by the jet stream creates a low pressure area whose rotating air distorts the interface (center) and finally creates the classic warm and cold fronts associated with low pressure systems (at right).

Pulled into the westerly circulation, the low pressure area moves eastward at 40 knots or more. If conditions are right, the system can develop further and eventually degenerate into a storm.

93. Hurricane, cyclone, typhoon ... is there any difference?

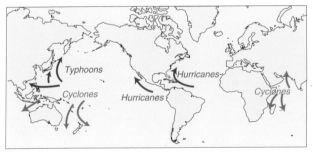

The hurricane's different names.

Americans call them *hurricanes*, a Carib Indian word that refers to the Mayan god of the heavens. For Europeans, the word is *cyclone*, a neologism based on the Greek for "circle." And for the same event in the seas of South-East Asia, we have *typhoon*, from the Chinese *t'ai feng*, "great wind."

But even if these powerful, traveling, rotating low pressure systems feared by all sailors in all the warm seas are called by different names, they are all exactly the same phenomenon.

94. How do hurricanes form?

Hurricanes only occur over warm seas and so are most frequent in the late summer of the hemisphere in question (September in the Northern Hemisphere, February-March in the Southern Hemisphere). The earliest stage is the *tropical depression* (tropical low), which forms when the following conditions are met.

- a water temperature above 80 °F to a depth of at least 150 feet (the massive evaporation of the water is what will provide the hurricane with its energy),
- a weak low at the surface, permitting convergence of the wind,
- a slightly anticyclonic zone at high altitudes, which permits upper-level winds to diverge and leave more quickly than surface air is converging (a hurricane works like a chimney, sucking the air up at the surface and expelling it at higher altitudes),
- not much variation in the winds between the surface and the top of the low-pressure system, so as not to "shear off" this chimney effect.

The process usually starts with thunderstorms that aggregate, then create a low-pressure rotational movement at the surface. If the process is amplified, the disturbance can become a *hurricane*. By convention, a tropical low becomes a *tropical storm* when the wind speed exceeds 33 knots, then becomes a hurricane when it exceeds 63 knots (12 on the Beaufort scale).

The same phenomenon cannot occur either over land or cold seas because there cannot be enough evaporation there to produce the necessary large zones with intense thunderstorm activity.

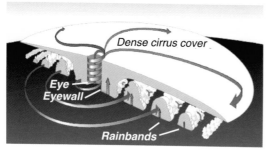

Section through a hurricane.

Air from the outer edges of the low that is sucked into the center acquires a spiraling motion thanks to the Coriolis force, turning clockwise in the Southern Hemisphere and counterclockwise in the Northern one. These winds, as they approach the center, are coming in from all directions and the only place they can go is upward. It is here, in the *eyewall*, with a diameter of about 30 miles, that the winds are the most violent and the rains are heaviest (up to one inch per hour). At the very center, in the eye itself, the air is calm. Actually, the air there is slowly descending. And it is dry, too, for it has lost all its water vapor, so the eye is clear and cloudless.

Hurricanes are enormous heat machines. If all the energy in a single one of average size could be converted into electricity, it would provide enough to satisfy the consumption of California for five full years! This very efficient way of transferring energy is what nature has invented for times when ordinary methods are not enough. For most of the year, the surplus energy we receive from the Sun in tropical zones is redirected towards cooler zones by ocean currents and the normal atmospheric circulation: trade winds, westerlies, etc. (Q. 88). By the end of summer, however, the equatorial and tropical zones are so hot that "peaceful" means no longer suffice. The safety valve on the pressure cooker must open. Hurricanes must be born.

95. When did the practice of naming hurricanes start?

We will all long remember *Agnes, Hugo, Katrina*, etc. Employing such short, easily remembered first names to identify hurricanes helps to make meteorologists' work easier and to sensitize the public. The practice began in the 19th century with the Spanish in the Americas. Hurricanes were referred to by the name of the saint on whose feast day the storm struck, *Santa Anna*, for example, for the hurricane that ravaged Puerto Rico in 1825 and *San Felipe* for the one that struck in 1876.

Early in the 20th century when meteorologists began following the paths of hurricanes, they identified them, not by name, but by the latitude and longitude at which they were first sighted. That system was obviously not very manageable for sending radio messages to ships. Beginning in 1953, American meteorologists began identifying hurricanes by the names in the phonetic alphabet then in use (Able, Baker, Charlie, etc.). But that led to some confusion when the International Phonetic Alphabet was adopted, and they then decided to use women's names. That continued to be the practice until 1979, when the Worldwide Meteorological Organization, under pressure from feminists, decided to use men's names, too.

For Atlantic storms, we now have six lists of first names in alphabetical order that are used in rotation, so the list used in 2006, for example, will be used again in 2012. Other regions of the globe have different lists using local names. In the Far East, where the idea of calling something as destructive as a hurricane by the name of a person is considered unseemly, the lists are made up of names of animals, flowers, rivers, etc., taken from the diverse languages of the region.

The lists of names are rarely changed, but there is a rule to the effect that, if a hurricane has been particularly destructive, its name is removed from the list.

96. Why is most hurricane activity concentrated in the western parts of oceans?

Once established, a tropical low moves westward, carried along by prevailing trade winds at a speed of about 10 knots. The low develops as it moves along, but needs several days to grow really big. This is why, although tropical storms are common enough in mid ocean, they usually only reach hurricane intensity in the western parts. There is one exception to this generality: since the Pacific Ocean is so large, hurricanes can develop in the warm waters off the coast of Mexico, then travel west to strike the Hawaiian Islands.

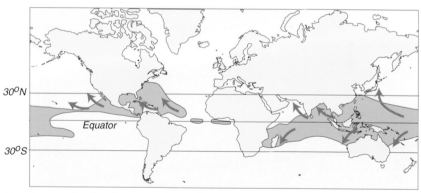

Hurricanes develop in areas where the water temperature exceeds 80°F (pink areas).

Carried along by the normal high-altitude circulation pattern, hurricanes usually first head west, then veer north in the Northern Hemisphere and south in the Southern Hemisphere, to circle around the subtropical highs, but some follow very erratic paths. Once they hit land they quickly die from lack of "nourishment" (in the form of water vapor) and from friction with the terrain. If they remain at sea, they lose their energy bit by bit as they pass over cooler seas in the mid latitudes.

97. Why are there no hurricanes in the South Atlantic?

The almost total lack of hurricanes in the South Atlantic is due to the low water temperatures prevailing there. The configuration of the continents, the wide opening onto the much colder Antarctic Ocean, and the pattern of the currents circulating there combine to prevent it from reaching the critical temperature of 80 °F over a great enough surface and to a great enough depth for hurricane formation. The warmest waters in the Atlantic are thus found, not at the equator, but just north of it. The Doldrums (also called the *InterTropical Convergence Zone* or ITCZ) are here, and so is the cradle of storm formation where tropical lows develop.

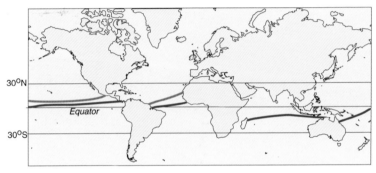

The Intertropical Convergence Zone in the Northern Hemisphere in summer (red) and in winter (blue).

The ITCZ is north of the equator in the eastern Pacific, too, and the waters off the coast of Columbia are hurricane free. But they do form off the coast of Mexico.[6] There are no hurricanes at the equator itself, either, not in any of the oceans. This is because the Coriolis force is zero there, and without it an air mass will not develop rotational movement. The Coriolis force can produce its effect only at latitudes of 5° (north or south) or higher.

98. Why does the wind abruptly change direction when a front passes?

The following figure is a classic weather map representation of a low. Zones of equal pressure indicated by the curved lines are called *isobars*. Note how they abruptly change direction as the front passes.

---

[6]The ITCZ is south of the equator in the Indian Ocean during the austral summer, however, because the Indian sub-continent blocks the northward transfer of equatorial heat.

This is due to the difference in temperature. Pressure at the cold side of a front is stronger than at the warm side because cold air is denser than warm air, and the front thus creates a discontinuity in the isobars.

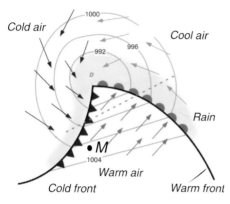

Since the wind blows almost parallel to the isobars (Q. 87), the abrupt change of pressure as the front passes is marked by an equally abrupt change in wind direction. If one is at point M, for example, where the wind is blowing from the southwest, it will abruptly shift to the northwest after the front passes. But the wind does not pass through the front since there is a mass of cold air on one side and a mass of warm air on the other. The front is thus actually a strongly turbulent zone separating northwest and southwest winds.

99. How do clouds and fog form?

Clouds and fog both form when a mass of damp air cools down. As the temperature drops, the molecules of water vapor suspended in the air mass become less agitated and condense to form tiny droplets of liquid water. If the droplets are very small, they fall slowly and the slightest updraft serves to keep them aloft. At high altitudes, these droplets form the clouds. If drops of water inside a cloud grow large enough, the updraft is unable to keep them from falling — it rains.

When water droplets form close to the ground, they create fog. Several different types of fog can form, depending on the cooling mechanism.

- When a mass of damp air encounters a cold surface (at sea, for example, with wind coming from the south, an air mass encounters ever colder water as it moves north), a persistent fog (called advection fog) forms and lasts until the wind drops or changes direction.

- When the land cools down during a calm, windless night, fog can form through contact with the ground (called radiation fog). This type of fog disappears as soon as the wind comes up or the sun burns it off.

100. Why do thunderstorms rarely occur over open seas?

Most thunderstorms (accompanied by lightning) occur over land; they are much rarer at sea. The requirements for developing one include, not only an unstable air mass (Q. 101), but also a little "push" from somewhere to initiate updrafts. On land, these little pushes are easy enough to come by: the ground, overheated by the Sun, increases the instability of the air mass, or a high landmass forces the air to rise. This cannot happen over the open sea.

Thunderstorms are the most frequent over large tropical islands and in warm continental areas where air masses with a high water content come in from the sea. This is the situation in the U.S. South, which regularly receives warm, humid air from the Gulf of Mexico, and also in Central Africa, where humid air arrives from the Gulf of Guinea.

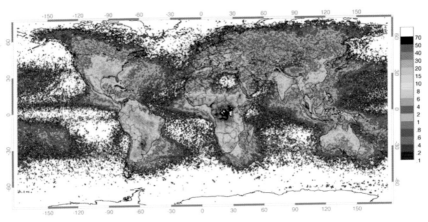

Average annual number of lightning strikes per km^2. Very few thunderstorms occur on the open seas, except in the Atlantic off the coast of the U.S. because of the warm waters of the Gulf Stream. (Data from 1995 to 2003, obtained from NASA's OTD and LIS satellites).

101. How do thunderstorms form?

Lightning is generated in enormous clouds, called cumulonimbus clouds, which form in humid, highly unstable air masses. An unstable air mass is one in which a layer of cold air sits on top of a mass of warm air (a situation that occurs, for example, when the ground has been strongly heated by the Sun and it, in turn, heats up the layer of air closest to it). The warm air in the lower layers, being lighter, rises.

And since pressure decreases rapidly with altitude, it then cools off (as a compressed gas expands, its temperature drops). At that point, the water vapor in it condenses into droplets to form a cloud.

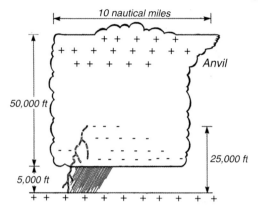

Lightning develops inside clouds at an altitude of about 25,000 ft.

Heat is liberated by the condensation of water vapor (Q. 90) and this warms the air, reinforcing its upward movement. The water in suspension, carried ever higher by the strong ascending air currents, is transformed into ice, a change of state which liberates even more energy, accelerating the whole process. Finally, an enormous cloud up to 10 miles thick, a cumulonimbus, has been created.

Why should lightning occur in clouds like these and not in others? The reason is their great size and the powerful air currents present inside them. In the high, cold regions of the cloud, the water freezes, forming ice crystals and hail. The exact mechanism that causes electric charges to appear is not yet completely understood, but it appears that collisions between the ice crystals and the hail balls are responsible.

During these collisions, the hail balls, which are a little warmer that the small crystals, acquire a negative electric charge[7] at the expense of the little crystals, which thus acquire a positive charge. The hail falls while the little crystals rise, carried upwards by the strong ascendant air currents in the cloud.

The upper part of the cloud thus becomes positively charged and the lower part negatively charged. The charges at the base of the cloud attract the positive charges that normally build up in the ground and, when the difference in potential between the ground and the base of the cloud is great enough, there is a violent electric discharge (the lightning bolt). This is exactly the same type of event as the spark in a car engine.

---

[7]These negative charges are OH^- ions. The H^+ ions tend to move towards the colder parts of the cloud and the OH^- ions towards the warmer parts.

The discharge has two components. The larger one originates inside the cloud and extends down to about 300 feet from the ground. There, it meets the second component rising up from the ground along a tree, a lightning rod, the mast of a boat, or other suitable object. The path of the lightning bolt looks quite wide to us but is actually no thicker than a pencil. The current is very high, however, about 30,000 amperes, and the temperature of the air it cuts through can reach nearly 35,000°F (3 times the temperature at the surface of the Sun). The sudden expansion of air creates a veritable explosion which, in turn, creates a shock wave over the first 30 feet or so, followed by the noise of thunder. Lightning travels through air almost as fast as light does (60,000 miles per second), so that the lightning flash lasts only a few millionths of a second. The noise of the thunder, on the other hand, lasts for several seconds since sound travels at a much slower speed (about 1000 ft/s). We first hear the noise caused by the part of the bolt that is closest to us (usually the part that hits the ground), followed by the noise of its passage through the successively higher layers of air. The thunder cannot be heard unless the lightning is less than 10 miles away. Beyond that, we can still see the flash, but the sound is too damped by air to be audible.

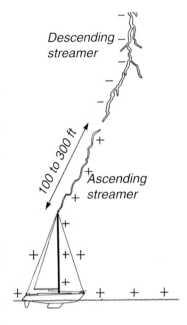

Ships

102. What is the largest ship ever built?

The Jahre Viking, a 1500-foot Norwegian oil tanker with a beam of 226 feet and a deadweight tonnage of 565,000 tons, takes the prize for the largest vessel ever built. Launched in 1979, she could transport 500,000 tons of oil but could only be accommodated in a few ports worldwide because, when fully laden, she sat 80 feet in the water; she could not pass through the English Channel or the Suez or Panama Canals nor enter most of world's major ports. She was converted to floating storage in 2004.

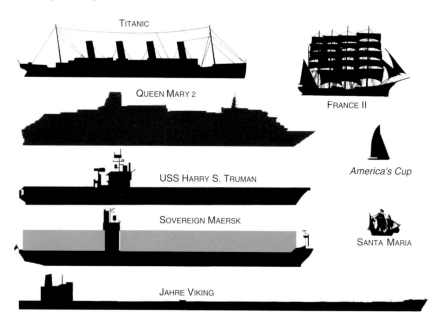

At left, silhouettes of the largest commercial and military ships. Shown for comparison at the same scale are the France II, *the largest sailing ship ever built, an America's Cup boat, and Christopher Columbus's* Santa Maria.

As for other categories, the largest cargo carrier is a 1,140-ft long Danish container ship, with a deadweight tonnage of 154,000 tons, the SOVEREIGN MAERSK. She can transport up to 6600 containers at a speed of 25 knots. The largest military vessel is the American nuclear powered aircraft carrier HARRY S. TRUMAN, 1,095 ft long, launched in 1997. She can carry 90 planes and 5000 men. The largest passenger liner is the QUEEN MARY 2, 1,131 ft long.

The honor of constructing the largest sailing ship ever built belongs to the French city of Bordeaux where, in 1911, the FRANCE II was launched. A five-masted ship 416 ft long, 5600 tons, she was constructed at this late date to transport nickel ore from New Caledonia to France. At the time, steamships could not make the long voyage because they were unable to carry enough coal for such a distance, and there were no convenient resupply ports en route. The FRANCE II sank in 1922 after hitting a reef near New Caledonia [41].

103. Over the past 4000 years of naval construction, how much has been gained in speed?

From ancient times until the 19th century, the most efficient way of moving men and materials over great distances was by sea. It was much more economical than transport by foot, cart, or camel caravan, and was often safer, too. Until the arrival of the airplane, it was also the speediest way, as can be seen in the records of ocean crossings listed below.

Records of ocean crossings

Date	Ship	Departure	Arrival	Time	Speed kts
Sailing Ships					
1846	YORKSHIRE	Liverpool	New York	16d	8.0
1853	NORTHERN LIGHT	San Francisco	Boston	76d	8.3
1854	FLYING CLOUD	New York	San Francisco	89d	7.1
1868	THERMOPYLAE	Liverpool	Melbourne	63d	8.0
1905	ATLANTIC	Sandy Hook	England	12d 4h	10.3
Steamships					
1838	GREAT WESTERN	Bristol	New York	15d	7.8
1934	BREMEN	Cherbourg	Ambrose Light	4d 14h	28.0
1937	NORMANDIE	New York	Southampton	3d 22h	31.4
1938	QUEEN MARY	Ambrose Light	Bishop's Rock	3d 20h	34.0
1952	UNITED STATES	Ambrose Light	Bishop's Rock	3d 10h	35.8

Although ships have improved greatly over the centuries in both carrying capacity and safety, they have improved surprisingly little in speed. The ancient galleys clipped along at 6 knots, the caravels of Christopher Columbus at 7 knots in a good wind, and few modern ships ever exceed 25 knots. The reason for this "sluggishness" is that ships must confront the large resistance factor inherent in moving through water, whereas the wheel on land benefits from the small value for rolling friction (particularly on rails), and a plane in the air encounters very little resistance to motion. So it is natural that much greater advances in speed have been made on land and in the air. What we see, if we simplify and smooth out the numbers a bit, is that there is a factor of ten between the speeds of each of these modes of transportation: cruising sailboat, 6 knots; car or train, 60 knots (~70 mph); plane, 600 knots — which brings us to that well known modern definition of sailing as "the slowest, most uncomfortable, most expensive way to get from one place to another."

104. How fast were the great clipper ships?

The clipper, as its name implies, is a fast-moving ship. In fact, it was the fastest type of commercial sailing ship ever constructed.

Developed in the United States between 1830 and 1860, clippers became the preferred ships for bringing immigrants to America, carrying fortune hunters to California and Australia during the gold rushes, and supporting the strong commercial expansion of the U.S. during that period.

The tea clippers TAEPING *and* ARIEL.

Clippers were designed for speed rather than carrying capacity. Normally three-masted with square sails, they were about 200 ft long and streamlined (length to beam ratio of five or six to one), with a long, slender bow and very tall masts with wide yards that allowed them to carry much more sail than other ships of their day. The captains commanding these vessels were often obsessed with speed, pushing their ships and crews to their limits.

At a time when merchant ships poked along at 4 or 5 knots, with their heavy hulls and round bows that had hardly changed since the era of the Spanish galleon, the American clippers were revolutionary. Fantastic long-distance racers, they could reach speeds of 20 knots, reducing to less than 100 days the long voyages that, until then, had required nearly a year.

It was a time for the making and breaking of records. In 1845, the RAINBOW made the New York - China round trip in seven and a half months. The JAMES BAINES crossed the Atlantic from Boston to Liverpool in 12 days and went around the world in 133 days. The FLYING CLOUD made the trip from New York to San Francisco via Cape Horn in 89 days, the NIGHTINGALE did Shanghai to London in 91 days, the SEA WITCH did Canton - New York in 81 days and the CHALLENGE, Hong-Kong - San Francisco in 33 days. In 1854, the LIGHTNING established the record for distance covered in 24 hours, 436 nautical miles, a record broken only long after the advent of motor-powered ships.

The "Baltimore clipper", the favorite ship of pirates and privateers early in the 19th century, was the precursor of the great clipper ships.

Why should the clipper have been born in the U.S., when the great naval power at that time was England? The answer here can be found in the shipyards of Baltimore on the Chesapeake Bay, which, since late in the 18th century, had been turning out small, swift ships, at first for coastal trade, then for Caribbean pirates and American privateers during the War of 1812 against England, and, finally, for the slave trade.[1] This experience with designing and building sleek, swift ships is what gave rise to the great clippers [12].

The clippers remained an Anglo-Saxon specialty: first America, then England and Canada. Not many were built in other countries, just a few in France, Germany, Denmark and Holland. America

---

[1]Slave-ships had to be swift in order to keep down the death toll affecting their wretched cargo and to escape the English warships that pursued them after England (in 1811) and most of the other major powers forbade the slave trade.

stopped building them after 1860 due to a drop in demand. Great Britain continued producing them for the tea trade until about 1870, but the opening of the Suez Canal in 1869 and the arrival of the motor-powered ship rang their death knell.

The famous schooner AMERICA, the first to win the racing cup that then took her name, was constructed along the lines of the New York pilot schooners whose sleek hulls resemble those of the Baltimore clippers.

105. What is the absolute speed record on water?

The absolute speed record on water, 275 knots, is held by Ken Warby on the SPIRIT OF AUSTRALIA, which was equipped with a jet

SPIRIT OF AUSTRALIA.

plane motor. The record-breaking event took place on a lake in Australia in 1978.

The record speed for sailboats, 46 knots over a distance of 500 meters, is held by three Australians on YELLOW PAGES ENDEAVOUR, a vessel with three planing hulls.

There was not all that much wind, barely 20 knots, on the day that record was set back in 1993. This boat requires winds of at least 15 knots to take off, but once planing she can go almost three times the speed of the wind. The previous record, a speed of 40 knots, was held by a French windsurfer.

As far as ocean crossings are concerned, there is Charlie Barr's mythical record set in 1905 on the schooner AT-LANTIC, a west-to-east Atlantic crossing[2] in 12 days. His record remained

YELLOW PAGES ENDEAV-OUR. *Photo by Philippe Schiller.*

unbroken for 75 years. Not only was Charlie Barr an excellent skipper (a three-times America's Cup winner), but he benefited from an exceptional series of strong depressions on his crossing. On the other hand, conditions aboard were rough. When one of the storms was

---

[2]From Sandy Hook Light Ship (near New York) to Point Lizard (England), a distance of 2925 nautical miles.

at its worst, the terrified owner appeared on deck and called for the sails to be reduced. Barr answered with "You hired me, sir, to win this race, and *by God*, that's what I am going to do," and promptly sent the man back to his cabin.

The first to beat Barr's record on the ATLANTIC was Eric Tabarly in 1980 on the hydrofoil trimaran PAUL RICARD, shortening Barr's time by almost 2 days, and new records have been set frequently since then. The current record for the crossing by a single-hulled boat has been held since 2001 by Bernard Stamm on the Open 60 ARMOR LUX with a time of 8 days, 1 hour, and for catamarans since 2001 by PLAYSTATION with the incredible time of under 5 days. The record for distance covered in 24 hours is 695 nautical miles, established in 2002 by the English catamaran MAIDEN 2 [87].

Two of the boats that have held records for the Atlantic crossing: the schooner ATLANTIC *in 1905 at left, and the catamaran* PLAYSTATION *in 2001 at right.*

106. When was the rudder invented?

A rudder mounted on a shaft seems such an obvious invention that we tend not to realize how recently it was introduced. Throughout antiquity, an oar was simply placed at the stern to serve the purpose. The true rudder on a shaft made its first appearance in China in the 1st century B.C., and seems to have been independently reinvented in Europe, probably in Holland, in the 12th century. Without this essential invention and one other Chinese invention that was just as important, the compass, Christopher Columbus and Vasco da Gama's long voyages of exploration would simply not have been possible.

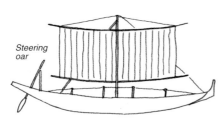 

At left, an Egyptian river transport boat from about 2400 B.C., with an oar mounted as a rudder. At right, the first representation (in Winchester Cathedral, England) of a rudder with stern-post, dating from 1180 A.D.

Until the mid 19th century, rudders were not "compensated,"[3] meaning that steering required the participation of several men and purchase-tackle (and later, a wheel). It was not until the famous English steamship GREAT BRITAIN, launched in 1843 (the first ship to cross the Atlantic ocean entirely under engine), that the compensated rudder appears.

107. Why was the lateen sail that can move upwind so well abandoned in favor of the square sail during the golden age of sailing?

The ancient western world knew only the square sail. It was used by the Egyptians to sail up the Nile in 3000 B.C., and by the Phoenicians, the Greeks, and the Romans. In the 6th century A.D., when the Arabs conquered the eastern and southern Mediterranean coastal lands, they imported naval techniques used in the Indian Ocean, including, specifically, the sail that came to be called the lateen sail. This is a triangular sail whose longest side is attached to a yard, called the *antenna*. The antenna acts as a stay for a jib, keeping the sail flat thus improving upwind sailing. A lateen sail can be used up to 50° from the wind whereas a square sail is limited to 70°. By the twelfth century, it had been adopted by virtually all seafarers in the Mediterranean,[4] in particular by the seamen of Genoa and Venice.

---

[3]Compensation consists of placing the shaft some distance behind the leading edge of the rudder so that the thrust of the water on the rudder area forward of the shaft at least partially balances the thrust on the area behind the shaft, thus reducing the torque needed to turn the rudder.

[4]It seems logical to assume that the name *lateen* refers to the use of this sail by people of the "Latin" world of the Mediterranean, but the term actually comes from the Italian *alla trina*, meaning "triangular."

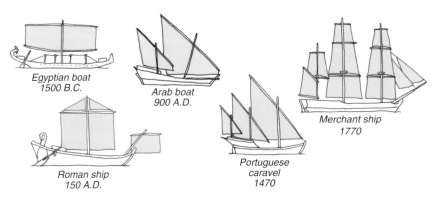

From the square sail to the lateen sail and back again.

The lateen sail, though efficient, has its drawbacks. The antenna is heavy and long, sometimes even longer than the boat itself, so that maneuvering it is difficult, and yet it has to change sides during tacking maneuvers to prevent the sail from striking the mast. That means that a large, experienced crew is required. Also, the sail cannot be reefed and must be lowered in strong winds, leaving the boat at the mercy of the seas. And finally, since the antenna sweeps over the entire bow, there is no way to install a stay, which limits the height of the mast supporting the sail.[5]

Clearly, then, the lateen sail is not well adapted to use on open seas or on large ships. The Portuguese used it on their caravels for the early voyages of exploration along the coasts of Africa and India: since they normally had downwind sailing conditions on the outbound leg, they wanted to be sure they could come back. But as soon as it became a question of ocean crossings and propelling large merchant ships, the square sail was back in favor. It was easy to handle and reef and also permitted the use of stays, which meant that masts could be taller and so carry more sail. Moreover, since winds on the open seas blow with more regularity than those near the coast, seafarers learned the best routes to follow to maximize downwind sailing, and the need to sail upwind became less pressing.

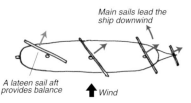

Still, one lateen sail was retained aft. This was a flat sail that could be sheeted in so as to counter the moment of the main sails and balance the ship at the helm. For the slow ships of the past, with their long keels and small rudder, that aft sail was also essential for tacking into the wind.

---

[5]The mast often had a forward rake to compensate for the lack of stay.

108. Where did the terms starboard and port come from?

Starboard is a term of Viking origin indicating the side of the boat that held the rudder oar, called a *styri*, a word related to the English verb "to steer." Since the (usually right-handed) person steering would be standing facing the bow, holding the styri with his right hand, this early form of rudder would naturally be mounted on the starboard side of the ship.

A Norman (formerly Viking) ship setting sail to take part in the invasion of England in 1066. Note the starboard rudder-oar. Detail of the Bayeux tapestry, with special permission from the city of Bayeux.

Now, when a ship was docking, this steering-oar on the starboard side had to be kept free to maneuver. The opposite side of the ship would therefore be the one brought up next to the dock. The early term for that side was *ladebord*, meaning the side from which the ship was laded. Over time, the term evolved, and *ladebord* became *larboard*. At that point, there was a possibility of confusion with the term *starboard* when orders were being shouted out, so a new term was needed. The choice fell on *porte*, French for "door," because there was, indeed, a door on that side of the ship, the door through which the cargo was carried aboard. And *voilà*, starboard and port [82].

Not surprisingly, a goodly number of sailing terms have Viking origins, including, besides starboard, "mast," "stay," "keel," and "luff."

109. When did sailors adopt hammocks for sleeping?

When Christopher Colum-
bus returned to Spain after
making first contact with the
Indians of the Bahamas, he
brought back some hammocks
and the Indian word for them,
hamaka. Actually, this word
simply meant "net" in the
language of the Arawaks and
Caribs, the indigenous peoples
that Columbus had encoun-

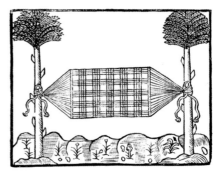

Carib Indian hammock.

tered. They used the article itself, made of palm or coconut fibers,
for fishing as well as for sleeping [55].

A hammock makes an extremely practical bed, cooled on all sides
and suspended out of reach of the crawling insects that abound in
warm climates. Sailors, especially the French and English sailors,
were quick to see the virtues of this economical, comfortable and
healthful sleeping arrangement. The only change they made was to
substitute canvas (cut from worn out sails) for the original net construction. An additional advantage of the hammock is that it sways;
if mounted parallel to the axis of a ship, the sleeper feels no rolling.
No more leeboards needed! Overall, it made for a much pleasanter
bed than most seamen were used to.

Rolled-up hammocks

*Hammocks stacked along the gunwale during mil-
itary engagements.*

But back in the
age of sailing ships,
providing for the
comfort of sailors
was the last thing
shipowners and
government agencies
had on their minds.
If the hammock
found favor with
them, it was for the
space it saved; once taken down, there was room on deck to set up
dining tables or to roll out the cannons. And during engagements,
the hammocks were rolled up and stacked along the gunwale to
protect the gunners from grape shot.

110. Why was the Viking tradition of lapstrake construction abandoned, when it produced such light, seaworthy boats?

When we see a dinghy or other small craft of lapstrake (clinker) construction today, we are moved to admire the elegance and lightness of this type of hull. It calls to mind the superb Viking drakkars, archetype of the seaworthy boat. Remember Erik the Red who sailed all the way to America four centuries before Columbus?

In lapstrake construction, the strakes (planks running stem to stern horizontally) are laid down first in such a way as to overlap, like tiles or shakes on a roof, and are then nailed to each other. The internal reinforcements (ribs) are installed next, and the strakes then nailed to the ribs. Because the strakes are nailed to each other,[6] the result is a light, flexible hull that remains watertight even when deformed by wave action.

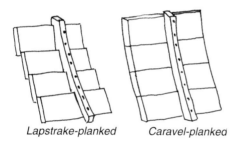

Lapstrake-planked   Caravel-planked

The other main type of wood construction, known as caravel construction, has been used since antiquity in the Mediterranean as well as in the Celtic world. Here, the internal framework is constructed first and then covered with the strakes, placed edge to edge. Once in the water, the planks swell up, compressing the joints, and any remaining gaps are calked. The resistance of this type of structure is entirely due to the framework, not at all to the planks. The result is a heavy, rigid, solid hull that can bear up under strong stresses such as those created by masts, stays, and rudders.

For vessels up to about 45 feet long, approximately the size of the drakkar, lapstrake construction is preferable because of weight considerations. But any larger than that and the hull is distorted too much when battered by the waves and becomes less watertight. Probably the largest lapstrake construction boat ever produced was the 200-foot GRÂCE DE DIEU, an English vessel launched in 1418. This type of construction was abandoned soon afterwards because only caravel construction can be successfully used in large, oceangoing ships.

In the late Middle Ages, just as shipbuilding and outfitting techniques on Europe's Atlantic coast had both strengths and limitations (lapstrake construction, square sails, true rudders), so did those of the Mediterranean world. The archetype there was the Venetian ship that

---

[6]Forming what structural engineers today call a "monocoque."

had sturdy, caravel construction but used lateen sails and had no true rudder (Q. 107). Ships such as these could not hope to cross oceans. Fortunately, though, thanks to increased contact and exchanges between the Atlantic and Mediterranean worlds, a cross-fertilization of naval construction techniques occurred soon thereafter and produced the hybrid, seaworthy ships needed for the great voyages of discovery, facilitating the rapid expansion of western civilization.

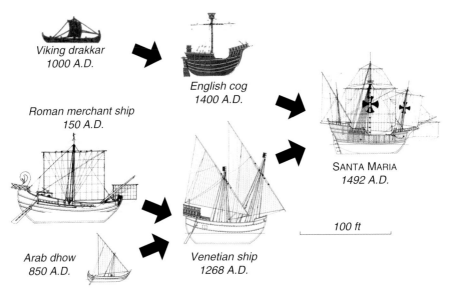

The European ocean-going ship of the 15th century is the result of a combination of Nordic and Mediterranean naval techniques.

The European ocean-going ship of the 15th century was thus a hybrid, combining the favorable characteristics of Nordic and Mediterranean ships: the southern caravel-planked hull and rigging with two or three masts, combined with the northern rudder with stern post (Q. 106), the square sail of the north on the foremasts and the lateen sail on the aft mast. After this symbiosis, which took place around 1460, the basic principles of European shipbuilding and outfitting remained almost unchanged for four centuries, until the appearance of steam power and steel.

111. Why is a ship's superstructure called a *castle*?

The term castle, designating the superstructure that shields the bridge on modern ships, is a legacy of the Middle Ages. At the time of the crusades, structures in the form of castle towers were constructed at the bow and stern of ships for the purpose of shielding the soldiers in case of attack at sea and allowing them to be higher than their assailants. These structures, which were temporary at first, became permanent fixtures in the 15th century, and the original name was simply retained.

112. What determines the size of a ship's wake angle?

Surprisingly enough, no matter what the size, shape or speed of a body moving through water, the angle of its wake is always the same: 39°.

Be it a duck, a sailboat or an aircraft carrier, the wake angle is always 39°.

The phenomenon was first explained in 1887 by the famous English physicist Lord Kelvin, and since then this particular angle has been called "the Kelvin angle."

The math is difficult and complex, but the solution can be envisioned geometrically [72, 23]. A ship's wake is the result of interference between waves that are continually produced as the vessel advances. It is made up of two different wave "families," one type transverse, forming on the stern at right angles to the direction of the

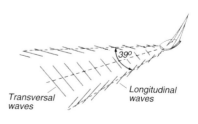

Transversal waves

Longitudinal waves

ship, and the other longitudinal, propagating parallel to the V of the bow wave. The multiple crests of the longitudinal waves give the wake a feathery look as opposed to the simple V that a supersonic plane produces. All the same, the case of a plane is the simplest, so we will look at it first. Just as a ship pushes the water in front of it aside, a plane pushes aside the molecules of air, creating pressure waves similar to waves on the surface of water. These pressure waves fan out in all directions, moving with the speed of sound in air.

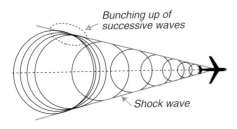

Bunching up of successive waves

Shock wave

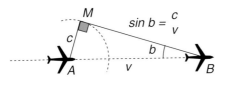

$$\sin b = \frac{c}{v}$$

When a plane's speed exceeds the speed of sound, the pressure waves "bunch up," compressing each other along an ever-widening cone and creating what is called a shock wave, which is simply an intensely loud sound wave. Just as a ship creates both a bow wave and a stern wave, a plane creates two shock waves, one at its nose and the other at its tail. These waves expand and, when they reach the ground, produce the well known supersonic boom.[7] The angle of the shock wave varies with the speed of the plane. The sine of the half-angle of the shock wave is equal to the ratio of the speed of sound to the speed of the plane. At Mach 2, for example, i.e. for a plane going twice the speed of sound, the full wake angle is 60°.

For a ship, the situation is similar but with one big difference: unlike pressure waves (sound waves), which all travel at the same speed, the speed of waves on water varies with the wavelength. When a ship advances, the bulge of water produced at the bow does not continue traveling as a single mass, but breaks down into individual waves, all with different wavelengths and therefore all moving at different speeds (Q. 14). The long waves, which move the fastest, quickly catch up with the shorter waves generated earlier, and when that happens their amplitudes (heights) are added together. When several such waves come together like this, the local amplitude increases, forming what is called a "wave packet." The original bulge of the bow wave thus moves away from a ship in the form of wave packets that are continually renewed as new waves form behind it.

---

[7]There are really two "booms," one caused by a plane's nose wave and the other by its tail wave, but most people cannot distinguish between them and just perceive the sound of a single explosion.

Toss a pebble into a pond and watch what happens: the crest of a wave inside the circles catches up with those ahead of it, then disappears when it reaches the outermost circle. The individual crests thus move faster than the wave packets. It can be demonstrated mathematically that the displacement speed of a wave packet in deep water is *one half* the average speed of the waves that constitute it at any given moment.[8]

We are now ready to look at what happens in the wake of a boat. Since the waves making up the bow wave travel at different speeds, the distance AM that we had in the example of the airplane now has a whole series of values. But since the angle at M is a right angle, all points of M are located on the circle with the diameter AB. Since the speed of the group is one half the individual speeds, the wave packet starting out from A will only have covered half the distance AM (at N for example). The locus of all these midpoints is the red circle, and all contributions to the wake when the ship was at A will be found inside it.

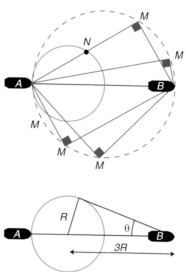

Since the radius R of this circle is one fourth the distance AB, the sine of the half-angle of the wake (θ) is equal to $R/3R = 1/3$, such that the half-angle of the wake is 19.5°, and the total angle is 39°. Note that, unlike the shock wave of an airplane, the wake angle here is *independent of the boat's speed*, at least as long as the boat is in deep water and not planing.

Since the contributing waves travel faster than the wave packets, we might be surprised to find no trace of them outside the wake. What happens is that, just as in the case of the pebble tossed into a pond, those waves die in the outermost crest of the wake. Further out, they are canceled by destructive interference.

Elementary, my dear Watson!

113. Where are the world's main fisheries found?

Fish play an important part in nourishing the 6 billion inhabitants of our planet. The average person consumes 30 pounds of fish per year, representing 16% of the total consumption of animal pro-

---

[8]In physics, we would say that the group velocity (the speed of the wave packet) is one half the phase velocity (the average speed of the individual waves).

tein. Most of this fish is wild-caught in the sea; freshwater fishing and aquaculture furnish only about one third of the total catch. And contrary to what you might think from the display at your local supermarket or fish store, the wild fish catch worldwide is dominated by the anchovy, the sardine, and the herring. These three species alone represent fully a third by weight of the annual world catch [20].

Unfortunately, this important natural resource has been badly misused; the ocean has been overfished since at least 1980. The codfish population that, for generations of fishermen, had seemed inexhaustible, has all but

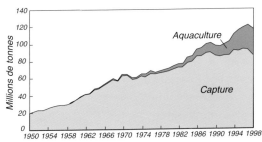

Annual worldwide production of caught fish and aquaculture. Source: FAO.

disappeared from Newfoundland and the North Sea. The anchovy fisheries off the coasts of Peru and South Africa have collapsed. All of the world's major fisheries are overexploited. It is estimated that the seas today hold only one tenth as many large fish as existed in 1950 [52]. At the current level of increase in world population, the consumption of fish per inhabitant is predicted to drop by 50% by the year 2020.

As the map below shows, most fish are caught on the continental shelves and in a few areas of upwelling (Q. 44).

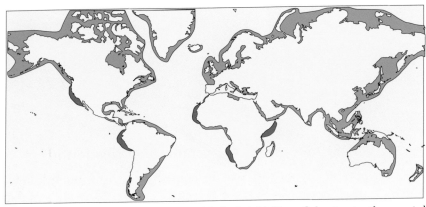

The major fishing zones of the world. In bright blue, fisheries in the coastal zones (continental shelves); in purple, the zones of upwelling.

114. Whatever induced European cod fishermen to sail all the way to Newfoundland for their fish?

In the Middle Ages, the Catholic church imposed a great many days of penance and abstinence during which the list of forbidden activities included making war, having sex and eating meat. All Fridays (the day of the crucifixion of Jesus) were meatless days, as were the forty weekdays of Lent and a whole series of additional special days on the religious calendar. Meat was thus banned from the table 160 days per year.

Meat was forbidden because it was considered to be a "warm" food that stirred the passions and excited the senses (leading to more war and sex). On the other hand, fish and water animals were authorized as being "cold" food.[9] This situation resulted in a heavy market demand for fish.

Unfortunately, it is hard to keep fish from going bad (Q. 205). In a world without refrigeration, one of the few ways of preserving food was by salting it.[10] Herring was preserved this way, but the flesh of herring is fatty and, even when salted and smoked, keeps poorly after exposure to air. Cod, on the other hand, has very lean flesh; once dried and salted, it can keep for years without turning rancid [39]. In the Middle Ages, this discovery was as important for people's diets as the sudden widespread availability of frozen foods has been in our own times. Codfish were abundant in the cold seas and were easy to catch, too, since they are found in relatively shallow water. Cod therefore became typical Friday food.

Salt was needed to preserve the cod, however, and northern countries bordering the cold North Sea do not produce salt.[11] The Basques and then the Portuguese, both of whom had salt, were thus destined to become Europe's main providers of salt cod. And since the great fishing grounds of the North Sea were already the domain of England and Scandinavia, the Basque and Portuguese fishing fleets were forced to work further away, in the waters off Newfoundland. It was far to go but the fishing was incredibly good. The schools of cod were said to be so dense that seagulls used them as rafts, to rest on, and

---

[9]Birds were considered to be cold food because they are not four-footed animals and, according to the Bible, were created on the fifth day, along with the fishes.

[10]Salt kills bacteria for the same reason that we cannot drink seawater: excess salt dehydrates living cells (Q. 40).

[11]Except for a few mines, the salt at that time came from evaporative "salt pans," and those require a hot climate.

dories sometimes sank under the weight of their catch after only half an hour of fishing.

The reason for such easily accessible abundance was the presence of the shoals, the famous "banks," particularly the Grand Banks, in waters rich with plankton, which is the basis of the food chain (Q. 56). All that crews had to do, once they had filled their holds with fish, was to go ashore at Newfoundland and spend a few weeks salting and drying the filets before sailing for home.

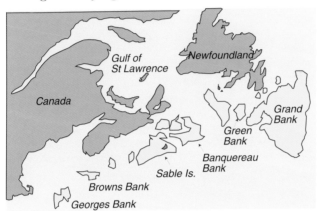

The "banks," shoals south of Newfoundland that abounded with fish and fed North America and Europe for four centuries.

After the Portuguese and the Basques, it was the turn of the Bretons, Americans, Canadians and many others to supply their fish markets with cod from the banks. But four centuries of unrestricted fishing there have taken their toll: the Atlantic cod population has essentially disappeared and does not seem to be recovering, not even after several years of a full moratorium on fishing (Q. 113).

115. Is fishing still a dangerous profession?

Fishing the oceans is one of the most dangerous jobs in the world, second only to professional logging. Mortal accidents in the fishing industry usually far exceed general national averages: in both the United States and Europe, for example, the number of such accidents is 20 to 30 times the national average. Out of 15 million men working on fishing boats worldwide, 24,000 accidental deaths occur every year [20].

116. Tonnage in tons, displacement in tons... are they the same thing?

Gross tonnage, net tonnage, Panama Canal tonnage, tons and tonnes, water tons, dry tons, and deadweight tons... There are indeed "tons" of such terms, and they can certainly be confusing. How to make sense of them, long and short, gross, metric, and all the others?

The word *ton* has a simple enough origin: the French word for barrel, *tonneau*, gave us the Old English *tunne*, which also meant a cask or barrel. Barrels were the "containers" of the past, watertight, stackable, almost unbreakable, and easy to move by rolling,[12] so they were ideal for transporting all kinds of merchandise on ships: water, beer and wine, salt fish and sauerkraut, gunpowder, iron nails, olives... For the English, who imported a great many casks of wine from Spain, evaluating the contents of the wine-bearing ships for tax purposes was easy enough: just count the "tunnes."

Now an ordinary barrel at that time held about 252 gallons of wine weighing about 2,240 pounds, so a *tunne* (soon to become *ton*[13]) could be used to indicate both the volume of the cask and its weight when full. Aye, there's the rub...and the beginning of the confusion. The ton became both a unit of volume and a unit of weight! And both volume and weight are used for measuring the size of a ship.

Clearly though, for cargo ships, what matters most is their internal volume for holding cargo. Volume units have evolved and have differed from country to country, but since the middle of the 19th century and under the influence of the British dominance of the seas, a ship's volume has traditionally been expressed in "tons" of 100 cubic feet ($2.83\,\mathrm{m}^3$). This is what is properly referred to as "tonnage." And since tonnage is a measure of the cargo capacity of a ship, it was also used to calculate government taxes, fees for docking and passing through canals, etc.

Depending on the application, one would use "gross" or "net" tonnage. Gross tonnage is the total internal volume of a vessel, with some exemptions for non-productive spaces such as crew quarters. Net tonnage is the volume available for cargo; it is equal to the gross

---

[12]Invented by the Gauls, who needed a container for their favorite beverage (no, not wine back then, but...beer!), the barrel replaced the amphora of the ancient world at the end of the Roman empire. A Gallic barrel held about 1000 liters of liquid.

[13]This is the "long ton," part of the Imperial system of weights and measures. It is close to the metric ton which is equal to 1000 kilograms (1 metric ton = 2,202 pounds) but should not be confused with the short ton used in the United States, which is defined as 2,000 pounds.

tonnage minus the volume of spaces that will not hold cargo (engine compartment, helm station, crew's quarters, etc.).

Since 1994, the "100 cubic foot ton" has been replaced by a new volume unit, the UMS (Universal Measurement System), whose exact value varies with the type of ship. The reason behind this new unit is that some types of vessels, such as container ships and roros,[14] carry part of their load above deck so that their cargo capacity is much larger than the strict volume of the hull. The table below gives the multiplying factor to be applied to tonnage expressed in traditional tons to find its new value in UMS.

Type of vessel	Gross tonnage	Net tonnage
Oil tankers	0.951	1.022
Bulk carriers	0.962	0.691
Cargo ships < 1600 tons	1.613	1.519
Cargo ships > 1600 tons	1.870	1.684
Roros	3.919	2.404
Car-ferries	1.599	1.260
Passenger ships	1.006	0.796

There is one category of ships for which cargo-carrying capacity is not essential, and that is the Navy ship. For these ships, total weight, which is indicative of the amount of their armor, guns and other military equipment, is more meaningful.

The total weight of a ship is called "displacement" because of Archimedes' law. As you may remember, this is the law of physics that states that the weight of a floating object is exactly equal to the weight of the water being displaced (i.e., that would otherwise occupy the "hole in the water" made by the object). Displacement is expressed in long tons or in metric tons (which have approximately the same value), and is calculated simply by multiplying the volume of the hull below the waterline (the volume of water being displaced) by the density of the water.

Weight, not volume, can also be useful in measuring the carrying capacity of merchant ships, as in the case of bulk carriers which transport materials of various densities. In this case, we speak of "deadweight." The deadweight is the displacement of a loaded vessel minus the lightship weight. It includes the crew, passengers, cargo, fuel, water, and stores. Like displacement, it is expressed in long tons or in metric tons.

---

[14]Roro is the acronym for "Roll On/Roll Off," a type of ferry or cargo ship that carries wheeled cargo such as cars, trucks, trailers or railway cars.

117. What kinds of engines are used on big ships?

The vast majority of commercial ships plying the seas have diesel engines. This type of engine has completely replaced the steam turbine except in a few special cases, such as methyl-transporters, where the evaporated gas is captured and used as fuel, in nuclear powered military ships, for reasons of range of action, and on ships that require a supply of steam for special uses (aircraft carrier catapults).

The diesel engine is simpler, more reliable, cheaper to buy and more economical to use. The efficiency of a steam turbine is only 35% while the diesel engine's is 50%. The diesels on commercial ships are generally 2-stroke engines revolving at about 100 rpm. The slow speed has two advantages. The first and probably most important one is that the slow piston movement permits the use of heavy oil fuel. Heavy oil fuel, a thick, blackish liquid, has to be heated before injection, but it contains more energy and is only half as expensive as the diesel fuel used in automobiles and trucks.

The second advantage of slow speed is that the engine can drive the propeller axle *directly* without requiring costly step-down gears. To increase the reliability and efficiency of the propulsion system, these engines have neither clutch nor reverse gear. To move backwards, the rotation of the engine itself is reversed. This is done using a two-position cam shaft that allows the valves and injectors to be retimed. The drawback here is that the procedure for reversing direction takes several minutes.

The largest diesel engine ever constructed for a ship (Sulzer RTA96C) during factory testing. The scale is given by the workman in yellow (circled).
Courtesy of Wärtsilä Corp.

Diesel engines on large ships are too massive to be "cranked" with electric starting motors; compressed air is sent directly into the cylinders to get them moving.

Engines on the largest ships are gigantic. The most powerful one, 100,000 horsepower, is currently produced by the Swiss company, Sulzer. It has 14 cylinders, is 80 ft long, 13 ft wide, and 40 ft high, weighs 2500 tons and consumes 3400 gallons of diesel fuel per hour! These engines are so big and heavy that they have to be delivered in pieces and assembled inside the ship during construction.

The current trend is to replace the "direct drive" diesel engines with more flexible *diesel-electrics*. In a diesel-electric, the diesel engine drives an electric generator, and the current produced is used by an electric motor to drive the propeller. That eliminates the need for the long propeller shaft of conventional systems and so cuts down on vibration. An additional advantage is that diesel-electrics can supply power for other needs, making them particularly attractive to military and cruise ships that are big consumers of electricity.

118. In emergency situations, what is the stopping distance of the largest ships?

The inertia of a commercial vessel is considerable. A supertanker cruising at full speed will cover several tens of miles after its engines are turned off before coming to a stop. Even with engines put into full reverse, a big tanker needs several miles to stop.

Generally speaking, the distance a ship travels before coming to a stop in an emergency (called a *crash stop distance*) depends on the vessel's displacement and speed, the power of its engine, and the time it takes to put the

	Crash stop distance (miles)
Car ferry	0.3
Liner	0.5
Cargo-ship	1.5
Supertanker	5

engine into reverse. This last factor can take up to several minutes for engines that drive the propeller directly (Q. 117). A few typical crash stop values are given in the above table.

119. Why is there a bulb on the bow of modern cargo ships?

Curious things, those large, bulbous masses visible on the bows of unladen cargo ships tied up at dock. The bulges are hollow and empty; their only purpose is to reduce the hydrodynamic resistance of the hull.

The gain in speed they provide was not deduced from scientific studies on the hydrodynamic behavior of ships' hulls. It was discovered quite by accident.

During World War II, a certain number of American warships were equipped with *sonar* to detect German submarines. Sonar, an acronym for "SOund NAvigation and Ranging," consists of emitting a sound signal in the water and listening for its returning echo. The position of the detected object is calculated from the direction of the echo and the time it takes to bounce back.[15] To prevent the noise of the engine, propellers, and bow wake from interfering with reception of the echo, the sonar equipment was installed inside a bulb in front of the bow. It was soon noticed that vessels thus equipped gained considerably in speed, or else, for the same speed, saw their fuel consumption decrease.[16] The fuel consumption is so much improved (10 to 15%) and can be obtained with so little expense that most merchant ships constructed after 1970 are equipped with bulbs, but minus the sonar equipment.

The explanation for the effect is that the water flowing around a bulb interferes with the bow wave, reducing its amplitude and, correspondingly, its resistance. A bulb's optimal size and shape depends on the speed of the ship, meaning that it is advantageous at only one speed, the ship's cruising speed. At slower speeds the bulb simply increases a ship's wet surface, producing a negative effect. Since great variations in speed are intrinsic to sailboating, and a bulb would be an advantage at only one of those speeds and a disadvantage at all others, a sailboat is definitely better off without one.

---

[15]Depth sounders use the same principle except that they give the distance in only one direction: straight down.

[16]This effect had actually been noticed earlier, in the late 19th century, when warships equipped with rams were being studied in testing tanks.

120. Are those enormous cruise ships with up to ten above-water decks really stable?

Those new cruise liners that sit twice as high in the water as they are wide are dedicated to transporting

Courtesy of Radisson Seven Seas Cruise Line

passengers. Hence, they have almost nothing in their holds to serve as ballast, and their centers of gravity are well above water level. How can they be stable in heavy seas or crosswinds?

Let us start by considering the case of a pendulum. It is evident that a pendulum is in unstable equilibrium if its weight is positioned higher than its axis. Even if we manage to balance it somehow, the least perturbation will make it

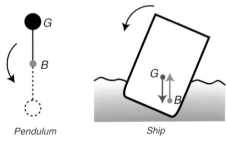

Pendulum   Ship

swing into its position of stable equilibrium, with the weight hanging down. The situation of a hull without ballast is somewhat different. The center of gravity (G) is certainly higher than the center of buoyancy (B), but the latter *shifts position* when the hull lists. If, as in the sketch, the hull lists to starboard, the submerged volume of the hull becomes greater to starboard, and the center of buoyancy moves starboard. The buoyancy force pushing upward and the weight of the ship pushing down produce a torque that tends to bring the ship upright again.

This is only true when the center of gravity is not too high, however. A great many ships have capsized because they carried too much weight too high up, either because they had excessive cargo stowed on deck, had a thick layer of ice on the superstructure, or because heavy waves were breaking on deck and were being evacuated too slowly. This is why oil tankers fill their tanks with seawater after delivering their cargos: unladen, they would sit too high in the water and be unstable.

The great height of today's large cruise liners, the QUEEN MARY 2 for example, with an air draft of 200 ft and a water draft of just 36 ft, is only possible because of their very wide beams and the use of modern materials to lighten the weight of their upper decks. Still, the

windage of these liners is so significant that it makes them vulnerable
to lateral winds and can handicap them during maneuvers in port.

121. How many large commercial ships are there in the world?

The worldwide commercial fleet is composed of about 90,000 ships
with a total capacity of about 600 million tons. This is three times
the number of commercial vessels that existed a century ago, and the
numbers continue to grow.

The merchant ma-
rine is the backbone of
world commerce, trans-
porting 95% of all mer-
chandise by weight.[17]
In tonnage, 60% of the
ships that ply the seas
are oil tankers and ore-

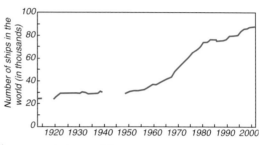

transport ships, while 20% are container ships.

One third of the world's fleet sails under what are called flags of
convenience (Liberian, Panamanian, Greek, Cypriot, etc.)[18] which,
although requiring compliance with international maritime conven-
tions, generally impose laxer rules regarding working conditions, na-
tionality, and payment of crews than those associated with the flags
of certain great powers. Next in percentage come Japan (10%), Russia (6%), and the U.S. (5%).

The transport of goods by sea is as successful as it is because
of its intrinsic low cost. Shipping costs per ton are only 2.5% of the
price of air transport.[19] That is due to:

- the "free" lifting force of water: as in the case of railways and
  roads but unlike that of airplanes, no energy is required to keep
  a ship afloat;

---

[17]Planes transport about one third of world commerce by value, however. Theoretically, air transport is only justified for light merchandise, valuable items, or
for objects needing rapid delivery (spare parts, for example). But in practice,
planes often serve as crutches to prop up poor business management techniques,
making up in delivery time for delays incurred in production time.

[18]With the absorption of Malta and Cyprus into the European Union, Europe
has become the largest maritime power in the world.

[19]But that does not hold true for small shipments or for passengers: sailing
across the Atlantic in one's own boat costs about 10 times the price of a plane
ticket!

– the small amount of energy required for displacement: the amount of energy required for movement is a function of the drag (friction), which itself varies as the square of the speed. In other words, the faster one goes, the more energy it takes to move the same mass. Now a plane has to go fast to stay aloft,[20] and a truck, too, has to be fast because the cost of a driver is high compared to the small quantity of goods transported. And although that is less true of transport by freight train, a ship is the least expensive method of all because neither does it need to go fast in order to float, nor is the cost of a crew large compared to the value of the cargo transported;

– the small cost of the required infrastructures: ports are necessary, but there is no need for roads, train tracks, or the vigilant attention of air controllers;

– the small cost of crews, thanks to the relatively low level of training they require.

Commercial ships, being "long-lived," are usually good investments for their owners: 58% of the world fleet is over 10 years of age, and 5% of it is 25 years old or older. But this intrinsic longevity can also be a factor in accidents when they are poorly maintained.

The seaworthiness of a ship is partly determined by the ratings given them by "classification societies." Among the oldest and most important of these is Lloyd's Register, which was first founded in 1760, then again in 1834; the American Bureau of Shipping (ABS), founded in 1862; and the Det Norske Veritas (Bureau Veritas), founded in Anvers in 1828, then moved to Paris in 1832. These societies publish rules that apply to the construction of the ships they classify and, subsequently, to the periodic inspections they require for a ship to retain its rating.

122. Where is the heaviest shipping traffic found?

Commercial shipping routes have always been determined by the existence of a market for the exchange of goods between two ports. The earliest sea routes necessarily hugged the coasts, but seafarers did not have to wait for radar and the GPS before striking out more boldly. Already by Roman times trading ships were sailing straight across the Mediterranean. When Europe began trading with the New World and the East over 400 years ago, the main factors influencing

---

[20]This would not hold true for dirigibles.

commercial sea routes were still the same as those affecting the ancient world: winds, weather, ice, and the topography of the planet. Winds ceased to be a major factor when steam engines replaced sails, and the planet's topography was modified for shipping when the Suez and Panama Canals were dug. Ship size has become a factor in that supertankers, which transport much of the world's oil, can only visit a small number of ports. But the surprise is that, despite advances in technology and the vast increase in world trade, commercial shipping routes have changed so little over the past four centuries.

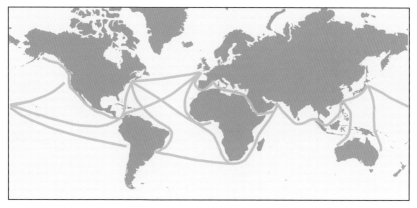

The main commercial shipping routes.

123. How big a crew do large commercial ships usually carry?

A large merchant ship usually has about 18 men aboard: the captain, 3 deck officers, 3 mechanics, 11 seamen and a cook. At night, an officer and a seaman stand watch. By day, the officer is usually alone on the bridge, but in case of need he can call on the members of the crew who are at their usual work stations (administration, maintenance, etc.). The captain does not usually stand watch. The seamen on watch are backed up by an automatic radar system (called ARPA for Automatic Radar Plotting Aid) that tracks the echoes coming from nearby ships and sets off an alarm when there is a risk of collision.

Commercial open-sea fishing ships usually have 4 to 6 men aboard: the captain, his first mate, the boatswain and his deckhands.

124. How many ships are lost at sea every year?

According to statistics published by the Institute of London Underwriters, the number of commercial ships over 500 tons lost due to collision, running aground, sinking, or serious accident (fires) is about one hundred per year. This represents an annual loss of 0.08% of the registered fleet by number and about half that by tonnage.

In times of war the statistics balloon. World War II, in particular, decimated shipping, merchant as well as military; the North Atlantic, particularly the English Channel, is littered with sunken wrecks. Between 1939 and 1945, a total of 5150 merchant ships were sunk, 2800 of them by Axis submarines, and to this we must add approximately 600 warships and 1000 submarines sunk in the Atlantic and Pacific (Germany alone lost 785 submarines).

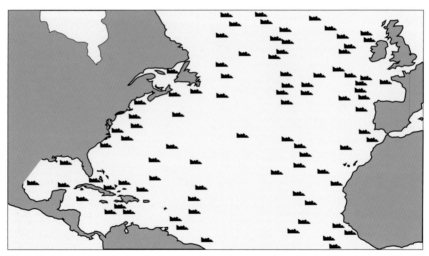

Merchant ships sunk by German submarines during the battle of the Atlantic between September 1939 and May 1943. Each symbol represents 10 ships sunk [74].

125. Do containers that are lost at sea represent a real danger?

At any given time, between 5 and 6 million containers are being transported at sea. Container ships, seeking to maximize their payloads, carry three quarters of their cargo on deck, with the containers stacked 6 or 7 high. In bad weather the lashings are subjected to great strain and, not surprisingly, can break.

Shipping companies do not make the reports of their losses at sea public, but it is estimated that between 2000 and 10,000 containers

The condition of the APL CHINA *on its arrival in Seattle in November 1998 after weathering a typhoon in the Pacific. She had lost more than 400 containers at sea.* Photo by Pat Brandow.

are lost annually [33, 73], which represents only 0.005% of the total. Still, the number of these "steel icebergs" left floating about is not negligible, particularly as most of them can stay afloat for months, being lighter than water even when loaded to the legal limit.

Containers are rarely truly watertight. If water leaks in slowly, say about 10 liters per hour, it takes approximately two months for a 20-foot container to sink, and about 6 months for a 40-foot container! Some containers will not sink even if water does penetrate because they are lined inside with thermal insulation (refrigerated containers) or contain electronic material (television sets, etc.) packed in polystyrene or foam rubber.

So lost containers do represent a potential danger, but estimating the real risk of collision is not easy. There have been several cases of sailboats colliding with UFO's (unidentified floating objects) over the past few years,[21] but in such dramatic circumstances a sailor tends to be more concerned about plugging up the hole in the hull or getting the liferaft into the water than about bringing out the binoculars to see what he has just crashed into. A container, maybe, but it could just as well have been a tree trunk or even a whale.

As a matter of fact, floating tree trunks probably represent more of a danger than containers do: they fall off loaded barges, break loose from floating log convoys, or are simply washed out to sea on

---

[21]Ellen MacArthur on her KINGFISHER during the Vendée Globe, Josh Hall during the BOC Challenge, Steve Callahan during the Mini-Transat [73].

river floodwaters. And a tree trunk can remain afloat for ten years or more. According to Japanese statistics, there are an average of 40 tree trunks floating around in every 10 x 10 mile square in the Pacific.

Still, the ocean is immense, and the chance of running into one of these floating dangers head on is small compared with the other risks incurred when sailing the seven seas.

126. How does a submarine dive?

The situation of a submarine floating at the surface is just the same as that of any ordinary ship: its buoyancy is greater than its weight, *ergo*, it floats. In order to dive, it has to increase its weight. This is accomplished by letting sea water into its water reservoirs, or *ballast tanks*:[22] valves in the upper part of the tanks are opened and the water comes in at the bottom, forcing out the air. When the weight of the submarine plus that of the water in the tanks becomes greater than the buoyancy force, the ship sinks, or "dives." As it goes down, the increasing pressure compresses the hull, reducing its volume and consequently increasing the force that pushes it down.

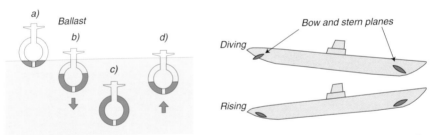

At left, the phases of a submarine dive: a) at the surface, ballast tanks nearly empty; b) preparing to dive, the valves are opened and water is admitted into the tanks, forcing out the air; c) once submerged, submarines use their fins to move up or down underwater as shown at right; d) to resurface, compressed air is used to empty the ballast tanks and lighten the ship.

On reaching the desired depth, neutral buoyancy is obtained by adjusting the amount of water in the submarine's *trimming tanks*, keeping its weight and the buoyancy force exactly balanced and allowing it to float, suspended, underwater, neither rising nor sinking. This is the same principle that fish use (Q. 42). Contrary to the

---

[22]Submarines have double hulls, a thick inner one designed to resist the enormous water pressure encountered during deep dives and a thinner outer hull containing the ballast tanks.

popular myth, submarines do not move up or down underwater by adjusting the amount of water in their ballast tanks. They control their depth by moving forward and adjusting the tilt of their fins (called "hydroplanes" or "bow and stern planes"), the way an airplane uses its wings.

To return to the surface, a submarine has to lighten itself so that the buoyancy force acting on it can overcome its weight. Compressed air is therefore pumped into the ballast tanks to force the water out, the ship becomes lighter, the buoyancy force becomes stronger, and the ship rises to the surface. It goes without saying that, before diving in a submarine, you should check to make sure that the supply of compressed air aboard is sufficient to bring you back up!

127. What is the purpose of a snorkel on a submarine?

Ordinary (i.e. non nuclear) submarines have electric motors running on DC current supplied by batteries. This allows them to be quiet and hard to detect underwater, but the batteries eventually need recharging. They are recharged by generators driven by diesel engines.

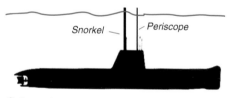

Submarine at periscope depth with snorkel raised.

When the ship is on the surface, the diesels have direct access to the air they need to function. If the ship dives and is to remain submerged, it moves up to "periscope depth," about 30 feet below the surface, and raises a tube to take in the air that its diesel engines require. This air-intake tube is the snorkel.[23] A flap at its upper end keeps water from running down inside when it is struck by waves. Exhaust gasses from the diesel engine are evacuated directly into the water so as to avoid detection. The submarine can then proceed to advance underwater with only the tips of its snorkel and periscope emerging above the water line.

The snorkel and periscope are all but invisible when the sea is agitated, but in calm seas the two tubes moving swiftly along at the surface of the ocean, leaving trails of foam behind, can be a strange

---

[23]The snorkel was a Dutch invention but was first used by the Germans during World War II, hence its German name schnorchel (schnarchen means "to snore"). Americans have simplified the word to "snorkel," while the British refer to the tube as a "snort."

sight indeed. Snorkels can be spotted by radar even in agitated seas and, as a consequence, are equipped with radar detectors so that, in wartime, the submarine can dive as soon as radar waves are picked up.

128. How are offshore oil platforms anchored?

When underwater oil deposits first began to be exploited back in the 1930's, an offshore platform was no more than a derrick standing in about thirty or forty feet of water. Things have changed a good deal since then, and we now find platforms hundreds of miles from our coasts, anchored in depths of 10,000 feet.

The deepwater platforms of today are floating structures held in position by multiple anchors or by cables attached to piles driven into the sea floor or attached to suction piles (SPAR platforms). Their great inertia coupled with their strong anchoring systems and the flexibility of their tethers allow them to survive even hurricanes.

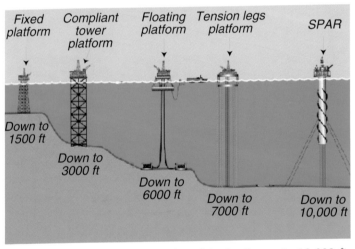

Offshore platforms can be anchored in depths up to 10,000 ft.

A new technique called "dynamic positioning" consists of providing a platform with motorized propellers linked to a GPS system that keeps it in position over the oil well below. This system, which does away with the need for anchorage, is mainly used for exploration and maintenance.

Yachting

129. When did yachting begin?

The *yacht*, i.e., a boat designed neither for transport nor war nor exploration nor display, but for the express purpose of sailing around for pleasure, is a Dutch invention of the 14th century. The Dutch had small, fast boats back then for chasing smugglers and pirates, boats they called "jaghtschippen," from "jagen" (to chase) and "schip" (boat). Rich shipowners eventually began using theirs to sail out to greet their merchant ships, just returned from the Indies. Later, it became fashionable to use the boats to take friends out for a little sail, simply for pleasure. Yachting was born.

A Dutch "jaght" from an engraving dated 1642.

It was the English king, Charles II, who transformed yachting into a sport. Charles was that Stuart king who, after spending 10 years in exile in Holland, was restored to the English throne in 1660. To celebrate the event, the inhabitants of the city of Amsterdam presented him with the MARY, a luxurious 60-foot "jaght" with a crew of 20 men. Charles took the MARY back to London with him, and with growing enthusiasm for his new possession, began to take pleasure in sailing up and down the Thames. He quickly developed expertise in navigation, in handling sails, and even in naval architecture, and had a good 20 or so "yachts" constructed (the word "jacht" had been quickly anglicized), several according to his own designs. Truly, he was the world's first "yachtsman." His brother James, Duke of York, also became an enthusiast, and soon the first regatta was scheduled on the Thames. It took place in 1661 over a round trip distance of 40 miles. On the starting line, KATHERINE, the king's newly-constructed boat, and beside her, ANNE, the Duke of York's yacht. KATHERINE crossed the finish line first, with Charles II, himself, at the helm. A new sport had been born.

MARY, *the Dutch yacht owned by Charles II of England, the world's first "yachtsman."*

The English yachts soon became the envy of all the crowned heads of Europe. The tsar of Russia, Peter the Great, also became an accomplished sailor. As for Louis XIV of France, he contented himself with commissioning half-sized replicas and had them sail around in the ornamental ponds of Versailles during galas. After the kings and dukes and tsars, the lesser nobility and rich men took up the activity. The first yacht club, the Cork Water Club, was founded in Ireland in 1720, followed in 1773 by two English clubs. In 1732, this charming definition appeared (in French) in Richelet's *Dictionnaire*:

> Iacht: the English or Dutch word which is pronounced "iac" in French. The English iacht is a vessel with masts and sail, appropriate for going to sea, made fine-looking with comfortable apartments, and pretty within and without.

The newly popular sport remained the prerogative of the British and a few Scandinavians for over a century. The first American yacht club, the famous New York Yacht Club, dates from 1844.

130. What are the major ocean races?

The first ocean race took place in the Atlantic in December 1866, departing New York. Three wealthy gentlemen from the New York

Yacht Club launched their schooners, VESTA, FLEETWING, and HEN-
RIETTA, on a race across the Atlantic, bound for the Isle of Wight,
England. Each owner bet $30,000, an enormous sum at the time, the
equivalent today of approximately 1 million dollars. The owners of
VESTA and FLEETWING placed their boats in the capable hands of
their crews and they, themselves, stayed at home, warm and safe.
But the third owner, James Gordon Benett, was aboard HENRIETTA
with her crew of 30 men. He won the race, too, after a 14 day crossing, and pocketed the $90,000. That race, which remained famous
under the name of "The Great Transatlantic Yacht Race of 1866,"
launched ocean racing. But it was not until early in the 20th century
that races began to be organized systematically.

Courtesy Christie's

*The start of
"The Great
Transat-
lantic
Yacht Race
of 1866,"
the first
ocean race.*

For the first two thirds of the 20th century, ocean racing was
dominated by four big classics:

The **Bermuda Race** — one of the two oldest ocean races along
with the Transpac. This is a handicap race that dates from 1906
and has been run in alternate years (even numbered) since 1924.
The course, from Newport, Rhode Island to Bermuda, covers a distance of 635 nautical miles and is enlivened by the crossing of the
Gulf Stream, with its powerful eddies. The location of the current is
usually determined by monitoring the water temperature, which can
reach 80°F.

The **Transpac** (Trans-Pacific Race) — it dates from 1906 and is
run every two years (odd numbered), alternating with the Bermuda
Race. This is a handicap race between Los Angeles, California and
Honolulu, Hawaii, a distance of 2225 nautical miles. It is usually a
very fast race, as the sailing is essentially downwind.

The **Fastnet** — a handicap race from Cowes (Isle of Wight) to the Fastnet rock south-west of Ireland, with a return to Plymouth. This is a difficult race over 605 nautical miles, first run in 1925. It was won that year by JOLIE BRISE, a pilot cutter, 45 feet at the waterline, built in Le Havre, France in 1913 and converted into a cruising sailboat. The cutter was bought by an Englishman, Commander Martin, who also won the Bermuda Race with her in 1926 and the Fastnet twice again, in 1929 and 1930. It was just after the first Fastnet that Commander Martin and a few others created the Royal Ocean Racing Club. An annual race at first, the Fastnet has been held in odd-numbered years since 1933, alternating with the Bermuda Race. It has been one of the components of the Admiral's Cup since 1957. Storms during the 1979 Fastnet race resulted in the deaths of 17 competitors and led to a major overhaul of the rules and the equipment required to compete.

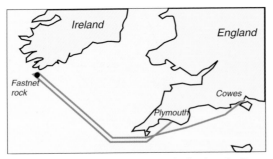 

The course of the Fastnet and the lovely JOLIE BRISE, *winner of the first Fastnet.* Photo: Beken of Cowes.

The **Sydney-Hobart** — a 630-nautical-mile race between Sydney, Australia, and Hobart, in Tasmania. It has been held annually at Christmas since 1945. The winds of the roaring 40's that come up as soon as you lose the protection of the Australian coast make this a hard race.

Besides the above, there are three important ocean competitions, each composed of several races:

The **Admiral's Cup**[1] — a tournament of nations with three boats per team, organized by the Royal Ocean Racing Club (RORC) and held biennially (odd-numbered years) since 1957. Each team participates in four races, two short and two long, including the Fastnet. This is one of the most prestigious prizes of ocean racing. However, after having to be canceled one year for lack of entrants, it is now

---

[1]The competition is now called the Champagne Mumm Admiral's Cup, as it is sponsored by the producers of the famous champagne.

looking to rejuvenate itself by changing the type of boat eligible for the race.

The **SORC** (Southern Ocean Racing Conference) — a series of six races, four triangular ones off the coast of Florida and two more, over 200 and 400 nautical miles, between Florida and the Bahamas. This is America's equivalent of the Admiral's Cup. As of 2005, the various SORC races are being organized by Premiere Racing.

Beginning in the 1970's, when the great classics of ocean racing had been in existence for over half a century, new races were added as a young generation of skippers and naval architects arrived, eager to face challenges to match their talents, ambitions and strengths. These are the "extreme" races where, not content to juggle winds, tides, and the pitfalls of the course intelligently, the racers must push themselves to the limits of human endurance and suffering. A large support team is often needed on land to manage logistics and "routing," and to find sponsors. These are impressive exploits, but one can't help regretting the passing of racing for the fun of it, a sport for enthusiastic amateurs, with cocktails at the club after crossing the finish line. It's a change that the venerable French magazine *Le Yacht* was already lamenting over a hundred years ago:

> "Yachtsmen will have to make up their minds if racing is to become once again an agreeable sport, or (remain) a fierce combat with pounds sterling for weapons and taking place in uncomfortable, if not completely uninhabitable, yacht-machines, constructed to function for only a few months..."

The major races of this now professional type of sailing are:

The **Whitbread**, now the **Volvo Ocean Race**: an around-the-world race in legs, with crews. After being run as a handicap race using IOR standards (1973-1994), it has now become a real time-on race on 70-foot boats built to satisfy a special rating, "70-ft Open" (previously WOR-60). The race takes place every 4 years, departing Southampton. The legs change from one race to another; in 2001 the stops were Cape Town, ports in Australia, New Zealand, and Brazil, Fort Lauderdale and Baltimore/Annapolis in the U.S., and La Rochelle, France. The race lasts approximately 120 days.

The **Around Alone**, (ex **BOC Challenge**): a solo around-the-world race with stops. Organized by the British Oxygen Company, the first of these races was the famously unsuccessful Golden Globe Race organized by the *Sunday Times* in 1968. Of the 10 participants only Knox-Johnston finished, Moitessier having decided to pursue his "long way." This new race is run in 4 or 5 legs every three to four years departing Newport, Rhode Island, with stops usually in Cape Town, Sydney and Rio de Janeiro, a distance of about 26,000 nautical miles. The obligatory stops have the effect of making this race even more stressful since the sailors, knowing that they can count on time to make repairs, push their boats harder than if they knew they were going to have to make it around the world non stop. The race takes about seven months and attracts the most prestigious racing professionals.

The **Vendée Globe**: a solo, around-the-world race, no stops and no outside help allowed, with start and finish at Sables d'Olonne, France. This is a pitiless race, considered by some as the toughest sports event of any kind.

The **Mini-Transat**: a solo transatlantic race on boats with a maximum length of 6,5 m. It is held every other year, lasts about a month, and is run between Concarneau (France) and the island of Guadeloupe in the Caribbean, with one stop in the Canary Islands.

The **Route du rhum**: a transatlantic solo race from Saint-Malo (France) to Guadeloupe (French Caribbean) that takes place every four years and lasts about two months. Single-hulled boats used to be able to compete, but the multihulls now have them outclassed.

131. How have sailboats evolved over the years?

When serious yachting began in the mid 19th century, naval architects naturally found their inspiration in the fastest professional ships of the day. In Europe, the model was the pilot cutter and, in the United States, the fishing schooners or pilot schooners of New England.

Pilot cutters had to be fast because pilots at that time competed with each other for business and worked on the principle of "catch as catch can." They would scrutinize the horizon and set out as soon as a ship came in sight; the first one to reach it would get the job. These boats also had to be very seaworthy, as they needed to go to sea in all kinds of weather, especially since, the worse the weather, the more an arriving ship needed a pilot's help.

An early 19th century English cutter.

New England fishing boats that went out to the "banks" (Q. 114) to fish also had to be fast. When the fishing part of the job was over, whoever got their catch back to port first would command the best prices.

At the time, marine hydrodynamics was still in its infancy, and it was thought that the best shape for a hull was the shape of a fish, with its big head and long, narrow tail. A wide bow kept the boat from plowing into the waves, and the slender stern calmed the wake. So the bow was full, the beam a third of the way from the bow, and the stern was narrow. This was the basic shape for merchant ships as well as war ships for several centuries. Things had to change.

In 1848, the great English naval architect John Scott Russell[2] called people's attention to the madness of these shapes and essentially proposed the reverse: a long, sleek bow associated with an ample stern [45].

Russell did not manage to overturn the convictions of English yachtsmen, but three years later, the AMERICA, constructed according to this principle by the New Yorker, Georges Steers, was destined to persuade them.

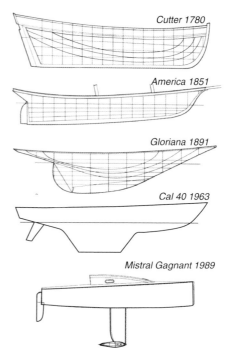

Cutter 1780

America 1851

Gloriana 1891

Cal 40 1963

Mistral Gagnant 1989

After the new shape was adopted, efforts were made over the next forty years to minimize the wetted area so as to reduce drag, but keels were still long and shallow. Evolution began with small yachts and found confirmation in GLORIANA, designed by Nat Herreshoff en 1891. Her deep, short keel designed to reduce friction and maximize the righting moment, together with her long overhangs forward and aft to lengthen the waterline when heeling, were revolutionary. The new lines were such an improvement that they would put their stamp on the next 70 years. But when carried to an extreme, they led to boats that were "skittish" (because the keel was too short), and not very maneuverable (because the rudder was too close to the center of lateral resistance).

---

[2]Scott Russell constructed the giant GREAT EASTERN. He is the author of a celebrated treatise on naval architecture and was recently designated the naval architect of the millennium by Great Britain's Royal Institution of Naval Architects.

The next step called for separating the rudder and the keel. That had already occurred on small boats (as early as 1890) but, strangely enough, it would be many years before larger craft adopted the separate rudder, It was even considered dangerous on ocean-going vessels. The first to defy this prejudice were Van de Stadt in Holland with ZEEVALK and BLACK SOO and John Illingworth in England with MOUSE OF MALHAM in 1954. But the *coup de grâce* was provided by the *Cal-40*, a mass-produced boat designed by Bill Lapworth in 1963. After several victories in the Transpac and the Bermuda Race, it became clear to everyone that this concept increased performance while retaining seaworthiness.

Forty years later, the principle of separating rudder and keel, with its variations, skeg on the rudder and bulb on the keel, remain the basis for modern hulls. The last time a long-keeled boat took part in the *Admiral's Cup* was in 1971.

The last step in hull evolution consisted of widening the stern to facilitate planing. Jean-Marie Finot introduced the idea in ocean racing with MISTRAL GAGNANT and THOM POUSSE, the great winners of the mini-Transat in 1989. Since then, this shape has been adopted for all high-performance ocean racing boats, such as the 60-foot Open.

Along with this evolution in hull shape, new, more homogeneous hull materials of molded wood, plywood, fiberglass and light alloys have led to considerably lighter hulls, limiting the need for the ribs of yore. The improvement in weight is so great that ocean-racing boats can now attain speeds up to double their hull speed, something never possible in the days of plank-on-rib construction.

So far, we have been looking only at the evolution of the traditional boat. Late in the 19th century, the "double boat" also made its appearance, with the first modern *catamaran* designed by Nathaniel Herreshoff en 1875. But the construction of multihulled boats only really began to flourish in 1960. Inspired by the Polynesian outrigger canoe, the catamaran,[3] the praho, and the trimaran evolved first as beach crafts, then for cruising and ocean racing under the impetus of passionate enthusiasts, often outside the mainstream of traditional naval architecture.

As for rigging and sails, there were two major steps in their evolution: (1) the adoption, around 1910, of Marconi rigging which increased sail efficiency thanks to a greater aspect ratio, and (2) start-

---

[3]Not a Polynesian word, contrary to what one might think (the Tahitian word is *pahi*). It comes from Tamil, a language of southern India in which *katu* means "tied" and *maran* means "tree trunk."

ing in 1950, the use of synthetic materials for sails, more durable than cotton or linen and which deform less under stress.

An important milestone in the evolution of yachting as a sport was reached early in the 1970's. That was when the racing yacht and the sailboat designed for Mr. John Q. Public went their separate ways. Until that time, ocean racing had been the domain of "amateurs" in the original sense of the word, devotees, lovers of the sport. Not paid professionals. Naturally, if you were just a pretty good helmsman steering a not-quite-brand-new boat, you did not expect to be among the first to cross the finish line. But it was still possible to *participate* for the sheer pleasure of it, on a modest budget and with a crew of friends. So almost all boats were designed as "racing cruisers," giving their owners a chance to race a little, if they were so inclined. After a few years, those boats became converted into just plain cruise boats, re-rigged for that purpose, but their lines and equipment were still influenced by racing and were designed for the corresponding ratings. The explosion of popularity for sailing in the 1960's led, on the one hand, to a general lack of interest on the part of buyers for racing-cruisers, which were deemed "elitist," and, on the other hand, to an out and out arms race for those who were determined to race. The upshot is that the sport is now split cleanly in two: pleasure boaters on one side and professionals or aspiring professionals on the other. The corresponding boats no longer have anything much in common. For the pleasure boater, comfort at sea and in port takes precedence over performance; for the professional, the reverse is true.

Some feel that this split between amateurs and professionals is a sign of decadence in the sport. The American naval architect Ted Brewer has the following to say about the current professional racing boat:

> I will not even grace them with the name "yachts" anymore because a yacht is a boat built for pleasure and there is not much pleasure in sailing aboard a modern ocean racer. I've been on ocean races where we sang sea chanteys on watch, had a happy hour in the late afternoon, roasts and pies at dinner, and a bottle of good wine to wash it down. We sailed for fun, and we won our share. That's pleasure, but I doubt if the today's owners and sailors get any true pleasure out of their sailing, unless they win!

132. Which are the greatest names in naval architecture?

We remember the names of a great many boats such as RANGER, MYTH OF MALHAM, PEN DUICK III, famous for having won prestigious races, or SPRAY or JOSHUA because they belonged to great navigators, or others, like *Swan*, *Cal-40* and *Figaro*, mass-produced boats that were huge successes with the boating public. But hardly anyone knows the names of their designers. That's a little unfair. If these boats are justly famous, it is not only because of who their owners were or which races they won, it is also because they were "good boats" designed by talented men, men who deserve to be recognized for the stamp they left on the naval architecture of pleasure craft.

The earliest yachts were designed by naval shipyards and were based on the boats they built for fishing or trade. That changed around 1890 when the problem was reconsidered in terms specific to pleasure boating: rating specifications, no cargo to transport, and, sometimes, with use limited to relatively calm waters. At that point, naval architecture for pleasure boats took off.

Three major periods can be discerned: the first, the era of the "founding fathers," running from the birth of yachting to 1920; the second, 1920 to 1960, the period during which technical aspects were mastered and the high point in design of individual boats was reached; and the third, the modern era, which has seen the "democratization" of the sport and the birth of sailing as a professional sport.

The founders — 1890 to 1920

Nathanael Herreshoff (1848–1938) — "Captain Nat Herreshoff," whose shipyard was in Bristol, near Newport, Rhode Island, was

Nat Herreshoff, GLORIANA, *and, at right, his extravagant* RELIANCE.
Photo: Rosenfeld.

incontestably the greatest American architect of his time. His designs were so innovative that he was called the "Wizard of Bristol."

We owe the first pleasure-boat catamaran to him, but his stamp has been the most indelible on the racing sailboat. His celebrated GLORIANA revolutionized the very concept of the racing boat, with her taut line from bow to keel (Q. 131). And he designed five America's Cup winners: VIGILANT (1894), DEFENDER (1895), COLUMBIA (1899 and 1901), RELIANCE (1903) and RESOLUTE (1920). Until 2003, RELIANCE was the largest single-masted sailboat ever constructed: 143 ft long, 88 ft at the waterline, 20 ft of draft. Her mast was 190 ft tall, the sail area 16,000 square feet (eight times as much sail as a modern America's Cup boat), and the mainsail sheet was 1000 ft long. It took a crew of 66 men to maneuver her during races! This gigantism marked the end of an era. Hulls and riggings had reached their limits of resistance, and these huge racing machines could only race around buoys in calm seas. Using them for cruising or ocean crossings was out of the question.

William Fife III (1857–1944) — If Herreshoff is the uncontested master of American yachting design at the end of the 19th century, the Scotsman, William Fife III, is his rival on the other side of the Atlantic.

His boats were less distinctive and less rapid than Herreshoff's (in the America's Cup, the first three SHAMROCK's that he designed for Lipton were never able to win over Herreshoff defenders), and he lacked his rival's scientific approach, but he is remembered as the greatest traditional naval architect.

William Fife and Éric Tabarly's PEN DUICK

The beauty of his lines and quality of his construction were unsurpassed.

Colin Archer (1832–1921) — In spite of his English-sounding name, Colin Archer was Norwegian, born of Scottish emigrant parents. He was famous for his sound, comfortable boats of all kinds. He designed FRAM for the polar explorers Nansen and Amundsen, and his pilot boats and lifeboats were particularly appreciated for their sturdiness and seaworthiness. He applied the same principles to his cruising yachts: comfort and seaworthiness, with a distinctive "Norwegian" (or canoe) stern. They were not racing machines, of course, but they made a lot of open-ocean sailors very happy. As proof of their appeal, new boats are still being constructed using his original plans.

Colin Archer and one of his boats with the Norwegian stern.

The classic masters — 1920 to 1960

Starling Burgess (1878–1947) — William Starling Burgess had a great model close to hand: his father Edward had designed three America's Cup winners in the 1880's. Fifty years later, it was the son's turn, and he, too, designed three winning J-class defenders, ENTERPRISE (1930), RAINBOW (1934), and RANGER (1937).

Starling Burgess with, to his left, Harold Vanderbilt, the owner of RANGER, the last and swiftest of the America's Cup J-class (center). At right, NIÑA during the Bermuda Race in 1956.   Photos: Rosenfeld Collection, Mystic Seaport.

His designs were perhaps not quite as elegant as those of his rival, Charles Nicholson, but they were faster. Their masts were aluminum, their decks equipped with winches (the British were still using block and tackle), and his hulls were studied in testing tanks, a first for yachting. Burgess also designed many racing and cruising yachts, including the lovely schooner NIÑA, with her spartan quarters and Bermuda rigging, that left her mark on ocean racing in the period between the two world wars. She won the 1928 Fastnet and, twenty-eight years later, cockily took first in the Bermuda Race. A brilliant and innovating architect, Burgess had wide interests: he published poetry, built airplanes under license to the Wright brothers, and legend has it that he created the printing style *Times New Roman*.

Charles Nicholson (1868–1954) — Nicholson was an admirer of Fife and strove to be his worthy disciple in the service of his patrons in the south of England. He was one of the first to use the Marconi rigging and designed four J-class challengers for the America's Cup: SHAMROCK IV, SHAMROCK V, ENDEAVOUR, and ENDEAVOUR II. ENDEAVOUR should have won the America's Cup, and the performances of his many 8 and 12 meter class remain first rate even by today's standards. His mastery of the distribution of hull volume is equally astonishing. The two ENDEAVOUR's, built in the 1930's, are considered to be among the most beautiful boats ever launched.

Between the two world wars, Camper & Nicholsons, the Nicholson family's shipyard based in Southampton, was the largest shipyard specializing in yacht construction in the world and it continues today to construct large sailboats.

Charles Nicholson at the helm of CANDIDA; *at right,* ENDEAVOUR II. *Photo: Beken of Cowes.*

Olin Stephens (born 1908) — The New Yorker Olin Stephens is without a doubt the most celebrated yacht architect of the 20th century. After leaving school at the age of nineteen, he went to work with a yacht broker, Drake Sparkman, and with him, in 1929, founded the company of Sparkman & Stephens that has since become world famous. Stephens remained the driving force in the company until his retirement in 1974. He created over 2000 designs during his 50-year professional career, and his boats were winners in all categories, from local regattas to the great ocean races, including the America's Cup.

Olin Stephens early in his career, his legendary yawl DORADE *and, at right, her equally famous successor,* STORMY WEATHER. *Photos: Sparkman & Stephens, Inc. (left) and Rosenfeld Collection, Mystic Seaport.*

The fundamentals around which he built his designs were a long, slim hull, low wetted area, a relatively deep keel, and smooth lines. This philosophy, which he applied from the very first, differed from that governing ocean sailing boats of his day, which were fairly wide and heavy and only sailed well downwind. As a trans-oceanic delivery captain ferrying meter-class racing boats, he had realized that the behavior and performances of such craft were perfectly well adapted to open sea conditions. He created a sensation when he won the transatlantic race of 1931, the Bermuda Race of 1932, and the Fastnet, twice, in 1931 and 1933, with his legendary DORADE, designed when he was just 22 years old. On returning to the U.S., he was accorded the highest tribute that America could then offer its heroes, a ticker-tape parade down Broadway.

Like his mentor Burgess, Stephens was convinced of the importance of tank testing and made generous use of it to refine the profiles of his America's Cup 12-meter racers. He designed four of them, winners all: CONSTELLATION in 1964, INTREPID in 1967 and 1970, COURAGEOUS in 1974 and 1977 and FREEDOM in 1980.

John Alden (1895–1962) — Extremely critical of early 20th century racing yachts for their lack of seaworthiness, the American, John Alden, designed a large number of highly appreciated ocean cruisers, mostly schooners.

Inspired by the New England fishing schooner, they were elegant boats, comfortable, designed for the high seas but without completely abandoning all thoughts of speed;

John Alden and MALABAR VIII, *built in 1927. Photos: Rosenfeld Collection, Mystic Seaport (left) and Alden Yachts (right).*

several of them did quite well in the Bermuda Race and the Fastnet. The seaworthiness of these boats was brilliantly demonstrated when SVAAP, the little 32-foot ketch that Alden had designed in 1925, circumnavigated the globe in the hands of the American navigator William Robinson.[4]

Philip Rhodes (1895–1974) — The American Philip Rhodes was a prolific architect, as "at home" designing little mass-produced boats as luxury yachts or minesweepers. Rhodes was even in on the America's Cup scene; he designed the 12-meter WEATHERLY that defeated Australia's GRETEL in 1958.

Philip Rhodes and CARINA, *one of his famous ocean racing yawls, winner of the Bermuda Race in 1952, the Cowes-Dinard in 1953, and placing second in the Fastnet that same year. Photos: Rosenfeld Collection, Mystic Seaport.*

[4]One of the first circumnavigations for a pleasure boat, but not quite a solo exploit as Robinson had a crewmember/cook along, a Bermudan at first, then a Tahitian.

In pleasure boating, he is primarily known for his graceful offshore centerboarders. At the time, centerboards were being dismissed as only good for "puddling around" in shallow, protected waters; only keeled boats could face the open sea and sail close to the wind. The many successes of his yachts in ocean events, such as the transatlantic races and the Bermuda Race, proved the opposite. Rhodes is about half-way between Alden and Stephens. His yachts were lighter and performed better than Alden's while being more spacious and more elegant than the narrow keelboats designed by Stephens.

Uffa Fox (1898-1972) — The Englishman, Uffa Fox, primarily designed small racing keelboats and centerboard dinghies, but he deserves to be mentioned here because he is the father of planing hulls in sailboats.

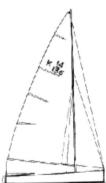

After working on motor racers, Fox was convinced that a centerboard sailboat could plane, too, provided that its hull was the right shape and that the crew kept it well upright. He applied this theory in his famous AVENGER, capturing numerous victories in the 14-foot international

Uffa Fox and his AVENGER, *the first planing dinghy.*

class. Just think...the very first planing sailboat. And it happened in 1928! Fox also designed large, classic yachts like WISHBONE, and left us many publications relating his own experiences, those of other architects, contemporaries of his and from the past, and yachting at Cowes.

John (Jack) Laurent Giles (1901–1969) — Over the years, the naval architecture firm founded by Jack Giles in 1927 has become the largest in Great Britain, rivaling Sparkman & Stephens. This firm has produced over 1000 designs, from racing and cruising sailboats to megayachts. Giles was known for his mid- to heavyweight ocean-going sailboats, particularly his famous VERTUE series, a 25-foot cutter of which over 230 were produced, crisscrossing the seas. He also designed WANDERER II, then WANDERER III, for the great seafarer Eric Hiscock.

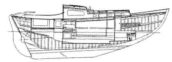

Jack Laurent Giles, his famous "pocket-sized" cruiser VERTUE, *and the revolutionary* MYTH OF MALHAM. *Courtesy of Laurent Giles Naval Architects Ltd.*

But Giles did not limit himself to solid, traditional sailboats. His MYTH OF MALHAM, designed for John Illingworth right after the war, was revolutionary for that time, with her weight-saving construction thanks to techniques borrowed from aeronautics, her narrow keel, shortened bow and stern, and mastheaded jib. MYTH OF MALHAM took the Fastnet in 1947 and 1949 and remained competitive in ocean racing for nearly ten years longer.

Giles was also the first to produce sailboats with reverse transom, which lightens the hull for a given total length, and with inverted sheer, which increases stiffness and inside volume. His MIRANDA IV, designed in 1951, had a rudder separate from the keel. With all these innovations, Giles announced the arrival of the modern period.

The moderns — 1960 to today

Bill Lapworth (1919-2006) — Bill Lapworth, from California, is responsible for designing the hull shape that is found everywhere today. His *Cal-40*'s, brought out in 1963, with their minimal wetted surface and rudder separate from the keel, was a sensation. CONQUIS-TADOR, the warhorse of the *Cal-40*'s, won the Transpac in 1965 and the two following years, and the SORC and the Bermuda Race in 1966.

Bill Lapworth and his Cal-40

The *Cal-40* was followed by a whole series of "Cal's," from 25 to 48 feet, all performing equally well. But Lapworth's real genius lay in wedding the concept of light weight with the advantages of mass production. Up until then, boats made of the new material called fiberglass were designed for coastal cruising. Few people thought they could perform as well in the great classic ocean races as custom-made boats costing three times as much. The *Cal-40* proved that a mass-produced boat could compete at the highest level, without even having been designed with ratings in mind.

It is true that sailing upwind in these light, flat-bottomed boats is not exactly restful. On the other hand, they take off and plane in a good wind, surfing at 15 knots for hours on end, a revelation for most sailors in those days who were convinced that exceeding the theoretical hull speed limit was impossible for an offshore boat. At such speeds, yawing is dangerous, but the long lever arm provided by the separate rudder brought the correcting power that was needed.

The Cal's were more than just racing machines. Beamy and spacious, they were wonderful cruising boats, and several of them made remarkable ocean crossings, precursors of what has become routine today.

Ericus Van de Stadt (1910–1999) — The Dutchman Ericus (Ricus) Van de Stadt is the pioneer of modern sailboat construction, first with plywood in the 1930's, then with fiberglass. His 9-meter PIONEER, which came out in 1955, was the first fiberglass boat to be constructed in series in Europe. In 1961, in collaboration with Illingworth and Laurent Giles, Van de Stadt designed the celebrated STORMVOGEL, a 22-meter ketch made of plywood.

Ericus Van de Stadt and his formidable "maxi" STORMVOGEL. *Courtesy of Van de Stadt Design.*

This lightweight yacht with the slim keel and separate rudder planed easily when the wind was up, and won practically every race she ran in, including the Sydney-Hobart.

André Mauric (1909–2003) — A racer, a builder, and an engineer all in one, the Frenchman André Mauric is one of the most prolific and creative naval architects of his generation. Straddling the classic and the modern eras, he began his career at a time when design was still based on half-hull models, but he evolved along with the new techniques and materials and produced truly fine, modern boats ranging from mass-produced sailboats to maxi-yachts and specialized craft.

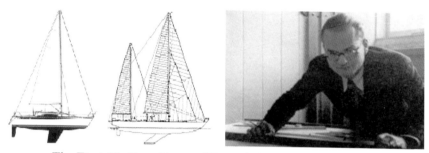

The First 30, PEN DUICK VI, *and André Mauric in 1927.*

He designed FRANCE 1 and 2 for Baron Bich's first French challenge in the America's Cup, KRITTER 5 and 8 for Michel Malinovski and Florence Arthaud, and 33 EXPORT, as well as PEN DUICK VI for Éric Tabarly. He is also responsible for the *First 30* series, with over 1000 produced and which served as the official entry for the sailing "Tour de France" between 1979 and 1981.

Germán Frers (born 1941) — The Argentinean Germán Frers learned his profession at Sparkman & Stephens before taking over his father's architectural firm in Buenos Aires in 1970.

Germán Frers, LUNA ROSSA, *winner of the Vuiton Cup in 2000, and the Swan 40.*

He produced more than 600 designs for clients the world over, from small sailboats to mega-yachts. His boats have collected the most coveted trophies, the Admiral's Cup, the Bermuda Race, the Transpac, the Giraglia, and the Whitbread race around the world. He also designed two Italian challengers for the America's Cup, MORO DI VENEZIA and LUNA ROSSA. Since 1979, he has been the official architect for the prestigious Finnish *Swans* series.[5]

Doug Peterson (born 1945) — An American from the West Coast, Doug Peterson stunned the sailing world when he won the One Ton Cup in 1973 with GANBARE, built with *U*-shaped sections at mid-length instead of the traditional *V*-shape. That bought him reduced displacement for the same rating and better performance when heeled. Besides which, with a shallower hull, the height of the keel was increased, improving upwind performance. That hull shape is now everywhere, even on cruising boats. GANBARE also marked a turn-around in the tendency to reduce sail surface so as to get a better rating. Racers had assumed that it was better to have less sail in windy courses in exchange for better ratings. Doug Peterson had given his boat a generous sail area to improve performance in light winds and downwind legs. After being a leader in ocean racing, Doug Peterson has now become the architect for the Italian challengers PRADA.

Bruce Farr (born 1949) — New Zealander Bruce Farr began his career in architecture working on racing dinghies, but he quickly moved up a rung or two on the ladder. His 45° SOUTH, designed when he was only 26, wasn't too impressive to look at, with her Sunday-afternoon-outing airs, and seemed to be the exact opposite of what the IOR rating had been set up for. But in 1975, to everyone's astonishment, she won the Quarter Ton Cup in Deauville. It was the first time that a boat designed and steered by New Zealanders had won a high-level international race. Farr did not stop there, of course, and his lightweight sailboats subsequently raked in so many first prizes that the rating authorities had to impose a 10% penalty on them to protect existing boats. Farr got even ten years later, however, when his boat took the Admiral's Cup with the New Zealand team.

Demand for his work is now so heavy in the Northern Hemisphere that he has transferred his office to Annapolis, Maryland, the sailing capital of America. Today, Farr is recognized as *the* specialist in high-

---

[5]The first *Swans* were designed by Sparkman & Stephens.

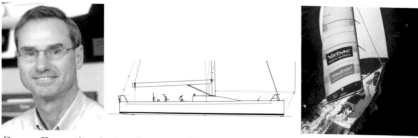

Bruce Farr, the design for one of his Transpac 52's, and for a 60-ft Open for the Vendée Globe. Courtesy of Farr Yacht Design, Ltd.

level races. His boats dominate all the races run on IMS ratings, and, since 1981, no around-the-world Whitbread has been run without his boats participating and often winning, as was the case five times in a row in 1986, 1990, 1994, 1998, and 2002.

Bénéteau, the giant of mass production, uses his services; over 800 boats in the 40–65 ft range have been built to his designs.

Farr even took a crack at the America's Cup when, in collaboration with Ron Holland, he designed the astonishing KZ7, the first fiberglass 12 meter, for the 1987 New Zealand challenge, then, in 1988, the gigantic KZ1 for the unequal race against Dennis Conner's catamaran STARS & STRIPES, and then YOUNG AMERICA in 2000.

Jean-Marie Finot (born 1941) — The Frenchman Jean-Marie Finot has been designing sailboats since 1964. His first success was the famous ÉCUME DE MER, which came out in 1968. This cruising sailboat, designed purely as an intellectual game with no thoughts of commercial marketing restraints, won a lot of races and established Finot as a first rate naval architect. Enlarging on his basic concept, he designed REVOLUTION, which won most of the class II RORC races several years running.

Jean-Marie Finot, the prototype for ÉCUME DE MER, FILA, *and Giovanni Soldini's 50-ft Open during the 1998–99 Around Alone.*

Since then, his racers have been scooping up prizes, particularly in the Open class. But Finot and his company (Groupe Finot) also turn out mass-produced boats, bringing to them the benefit of their experience with racers. Jean-Marie Finot is also the designer of the *Figaro* series, a monotype for solo races designed in 1989 in collaboration with Jean Berret. A total of over 30,000 cruising sailboats are out there now sailing on Groupe Finot designs, most of them constructed in the U.S., France, Italy, and Japan.

Gilles Ollier (born 1948) — The Frenchman Gilles Ollier is *the* specialist of ocean-going multi-hulled boats. He designed most of the famous racing catamarans, including JET SERVICES I & II, and 33 EXPORT. COMMODORE EXPLORER, the first winner of the *Jules Verne Trophy*, was also one of his designs, as were the four competitors in *The Race 2000/2001* that pocketed the first places.

The theoreticians

Naval architecture is an art — in the case of recreational boating, the art of designing boats that correspond to specific criteria of use while still remaining seaworthy, pleasant to steer, comfortable, and aesthetically satisfying. But science plays a big part in the process, too. Instead of trial and error, hunting for new solutions at random, science lets a designer zero in on a problem and predict the performance and behavior of the finished vessel. In this, recreational boating owes a great deal to three fine theoreticians:

William Froude (1810–1879) — An English engineer and naval architect, William Froude was the first to understand that the resistance a boat encounters to its progress depends upon two independent factors: friction on the submerged surface of the hull and wave-making resistance (Q. 138). He also discovered the laws that allow the performance of a ship to be predicted by scale model testing. And his analysis of ship movements remains the basis for the stability studies that are done today.

Manfred Curry (1899–1953) — An American who lived in Germany, Manfred Curry did the first scientific study of sail aerodynamics back in the 1920's. At the time, of course, aviation, though still in its infancy, had already produced an understanding of the main principles. But Curry applied these principles to the particular case of sails and did numerous wind tunnel tests to confirm his results. We are particularly indebted to him for the understanding that the negative pressure on the back of a sail has a much greater effect than the pressure on the wind side. His book *The Aerodynamics of Sails and Racing* is a classic.

Czeslaw A. (Tony) Marchaj — A former Polish Finn-class champion transplanted to England, Czeslaw Marchaj has become a specialist in the hydrodynamics and aerodynamics of sailing. He has done a great many tests in wind tunnels and testing tanks, and has made his accumulated expertise available to all in a number of seminal books, notably the *Aero-hydrodynamics of Sailing*.

133. What were half-hull models used for?

Who among us has never spent a few moments in quiet admiration before a half-hull model, one of those pretty scale models of the hull of a boat cut lengthwise, the dark wood carefully sanded and varnished, the whole nicely framed and (probably, now) adorning the walls of some yacht club? Just art objects for us today, these models were essential tools for naval architects and boatbuilders back in the days before computers.

A hull has a subtle, three-dimensional shape that is not easy to represent in drawings, so the architect used to sculpt the shape out of wood, modifying it until he had what he wanted. Since hulls are symmetrical, one half was all he needed. He would show it to the future owner for approval, then the profile would be "read off" by means of a pantograph (or, more simply, by sawing the model into slices to follow the contours). Once the boat was built, the half-hull would usually be presented to the owner, who would proudly hang it on his living room wall or present it to his yacht club as a gift.

This method for transferring the three-dimensional hull shape onto two-dimensional paper plans was slowly abandoned in favor of a drawing-table procedure in which long, flexible wood or metal strips, held in place against the paper by a system of weights and hooks, were bent to follow the curves, smoothing out the shape between the intermediate measured points. Today, the lines of the hull are balanced and smoothed out by computer, and being able to visualize

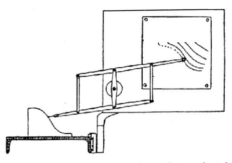

At left, half-hull models in Nat Herreshoff's study in 1938. At right, a sketch showing how profiles of the models were read with a pantograph.

the results on the screen means that models are no longer needed. But who wants to caress a computer screen...

134. Why do certain sailboats seem particularly beautiful and not others?

The poet Baudelaire once said that for him, the three most beautiful things in the world were a woman dancing, a horse galloping, and a ship under sail. It is true that certain boats cutting gracefully through the waves, white sails "big bellied with the wanton wind" (Shakespeare), can fill one with admiration. But it is just as true that other watercraft leave us pretty much unmoved, looking no more poetic than floating tree trunks. And painters no longer seem so eager to paint our modern marinas, where once the wooden boats in fishing ports were their favorite sources of inspiration.

Can we say that certain rules govern the aesthetics of boats, just as certain rules do for the composition of a painting or a piece of music?

First, we have to remember that our ideas of beauty change over time. Until the 19th century, boats were not considered to have any intrinsic beauty at all. To make them beautiful, they were painted, gilded and carved, and had statues and other adornments affixed to them.

These decorations were supposed to make one forget a boat's intrinsic lack of aesthetic interest, while at the same time symbolizing the owner's power and wealth: they added majesty, they impressed. This was particularly true of royal galleys and warships.

In modern times, though, we tend to feel that beauty can be found in most objects on condition that their forms respect the laws of nature. So we like simplicity, pure, uncluttered, unadorned lines that let an object's function shine through.

This way of looking at things is probably justified for an airplane, whose visual aspect is mainly determined by aerodynamics, and maybe, too, for a well-designed fork that balances nicely in the hand, but it would be a mistake to think that a boat's beauty stems from its functional virtues. We have to give up the myth that a *"beautiful* boat is necessarily a *good* boat"; beauty has little or nothing to do with a boat's seaworthiness. Some truly ugly boats are fast and sail like dreams, just as some with gorgeous lines are duds as far as handling goes or couldn't leave the harbor without falling to pieces. A good boat is good because of her hydro- and aerodynamic properties and the quality of her construction. Beauty and elegance, if present, are bonuses.

If that is the case, why do we feel pleasure when we see an old sailing ship or a sleek America's Cup boat? Let's remember, first, that beauty is not something absolute and objective, but entirely subjective and strongly influenced by our culture.[6]

Our judgment is influenced by our habits, desires, imagination. We marvel at old rigging or a wooden hull out of nostalgia for the age of sailing ships. But, as westerners, we could not feel quite the same way about Chinese junks, for example, because we know them and their history less well. Some people prefer a good, strong hull because it reminds them of sturdy fishing boats, others swear by the modern yacht with the high-tech look because it makes them think of the professional racers they adore, and still others love the long, graceful lines of a Fife or a Herreshoff because she looks fleet and calls to

---

[6]The philosopher Immanuel Kant was the first to propose the idea that aesthetic appreciation is not the perception of intrinsic beauty, but is due to our subjective judgment. Also according to Kant, the more our imagination is "informed," the greater is our pleasure. In other words, aesthetic appreciation has to be developed.

mind feminine curves. From a masculine point of view, in any case, a boat can have near erotic connotations. Doesn't the sailor love her? Doesn't he go off alone with her? Don't we refer to "her" as a "she" (the only inanimate object in the entire English language to be so personified)?

Proof that tastes change: BRITANNIA, *the British royal racing yacht built in 1893, which today we would find magnificent, was judged to be hideous in its day [27].*

If culture and the times affect our aesthetic judgment, there remain, all the same, a certain number of basic values that seem to be universally appreciated. Very generally speaking, beauty is harmony. Incoherence, lack of unity, are perceived as ugly.

Using this as a basis, the English architect D. Phillips-Birt set down a few principles of sailboat aesthetics half a century ago, principles that remain valid today [60].

- When a deck line is interrupted by the cabin superstructure, the harmony of the boat is lost. Hence the idea of integrating the cabin with the deck, eliminating as far as possible strong lines marking their separation.
- Continuous lines and repeated motifs give visual unity to a boat, as illustrated by the effect of multiple sails and the line of portholes along the side of the hull.
- The coachroof line should harmonize with the deck by being parallel to the sheerline or in line with the top of the bow.
- An elongated hull suggests speed but also fragility, while a "stubby" hull gives the impression of sturdiness and power. This is especially true if the curve of the sheerline is reversed, as in modern motorboats.
- Overly simple geometric shapes like the square, triangle, straight line and segment of a circle are as "boring" in boats as

they are in monuments and paintings. The eye finds more pleasure in shapes with changing curves and modulated surfaces. A hull with fluid lines and carrying multiple sails is enhanced by a lovely play of light and is pleasanter to look at than a stamped-out hull topped with flat, triangular sails.

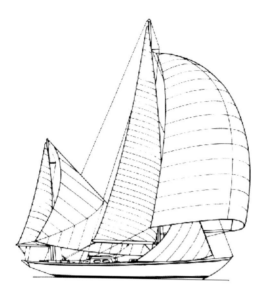

The multiple curves of this yawl's sails create a harmonious composition.

- The hull lines fore and aft are important to the general aesthetics and must be harmonious. Overly predictable straight lines and arcs of circles should be avoided; continually changing curves should blend fluidly with the lines of the hull itself.
- A normal sheerline rising towards the stern coupled with a classic overhang is graceful and, in a sense, spirelike, while the reverse curve sheer with no overhang can lead to an unpleasant termination aft. This is where an inverted transom can help by softening the abrupt sheerline end.
- Matched colors unify; gaudy colors are jarring, as a racing boat completely plastered with the logos of its multiple sponsors clearly illustrates.

But the rules that govern art do not create art, and the genius of a great naval architect is to transcend these rules and make the sailboat into something that is very like a work of art.

Besides these general features that can be judged from a distance, what we see on closer view obviously counts, too: the floorboards of a teak deck, the shining brass, the interior details, the small, discrete pleasures of touch and smell (wood is unbeatable here). And finally,

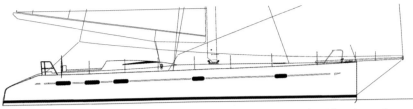

The elegant lines of a modern boat (here, a Finot design) without overhangs and with integrated deck and roof. The reverse transom harmonizes nicely with the slightly reversed sheerline.

on a boat, order is indispensable — and not only for the aesthetics! At sea, if everything is not meticulously stowed and lashed down, the swell will have everything topsy-turvy in no time, not only making the boat unsightly to look at but also difficult to manage in bad weather. And in the harbor, too, neatness counts: an impeccably organized boat tied up at the dock always attracts admiration.

VANITY 2, *the beauty of wood (restoration by Ribadeau-Dumas).*

Today's recreational boats designed for the mass market have to satisfy many conditions that conflict with aesthetics. Ratings, marinas' size limitations and price considerations tend to result in reduced or absent overhangs. In the search for comfort, roofs are raised excessively. Safety concerns impose high freeboards. It's a miracle that even a few boats still manage to look elegant.

Since the prime consideration in this sector is to turn out fast, seaworthy, comfortable boats at competitive prices, the product is essentially designed to meet marketing criteria. To this end, an architect in charge of the purely technical aspects is often teamed with a "designer" whose mission it is to design the exterior lines and, most importantly, an interior attractive to buyers.

135. Were those elegant old sailing yachts really seaworthy?

With their superb hull lines and gigantic sails, racing yachts of the second half of the 19th and early part of the 20th centuries are generally admired by all. Were they nothing but racing machines, devoid of true seaworthiness?

In England, as of 1855, racing yachts were designed for handicap racing using the Thames rating

$$\frac{(L - B) \times B \times D}{94}$$

where L is the length of the boat, B the beam, et D the depth of the hull estimated at $B/2$ so as to avoid having to haul out the boat to measure it. All measurements were in feet, and the formula gave the rating in English tons. This formula was inspired by merchant marine ratings in use since the 17th century. It actually gives an approximate interior volume for a hull, and consequently, a boat's payload. But its application to racing yachts quickly led to aberrations.

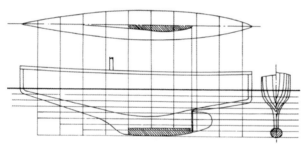

The cutter EVOLUTION, built in 1880, 50 feet long, 10 feet of draft and under 6 feet of beam is the absurd result of the Thames rating.

The fact that length was measured between the bow and the axis of the rudder encouraged a straight bow along with an exaggerated overhang aft to increase the waterline while heeling. Since sail surface was not taken into account, yachts became wildly overcanvassed. The formula also led to an increase in draft (which was not counted) and a decrease in beam (which was penalized). What resulted were ridiculously narrow hulls (beam/length ratio of 1:8, compared to the 1:3 of today's sailboats). These overcanvassed "planks-on-edge," as they were called, were dangerously unstable, and that rating formula had to be abandoned in 1887 after OONA sank with the loss of all hands when her hull split open under the action of the keel.

In contrast, American yachts were beamy and had low drafts like the merchant and fishing boats that plied the East Coast of the U.S. The generally shallow, calmer waters there (in summer, at least) favored this type of boat. But excess was to be found there, too. In

1876, the MOHAWK, a 140-foot yacht, beamy but with only a 6-foot draft, capsized in a strong gust of wind while setting anchor with all sails still up in the very port of New York.

Eventually, reason prevailed and rating formulas were changed to take stability into account. The very end of the 19th and the early 20th centuries saw many excellent boats being built, both comfortable and seaworthy. This was particularly true of European boats, which sometimes had to face worse weather during the summer yachting season than those encountered off the American coasts [46].

But seaworthiness sometimes came with a high price. For years, English challengers for the America's Cup had a major disadvantage to overcome. The rules of the race specified that the challengers had to come to the site of the race "on their own bottom," as AMERICA had done. For U.S. entries, that meant a short trip in the protected waters of Long Island Sound, so the boats could be lightly built. But the English challengers had to start by crossing the Atlantic...

136. Are sails the only way to harness wind power to propel a boat?

In 1926, a most unusual boat crossed the Atlantic: the BADEN BADEN, a former schooner whose rigging had been replaced by two rotating cylinders. The invention was the brainchild of a German aerodynamics expert, Anton Flettner.[7]

The Flettner rotor ship later renamed the BADEN BADEN.

---

[7]Anton Flettner (1885-1961) is also known for constructing one of the first helicopters early in the 1940's, and for his invention of the trim tab for airplanes.

Flettner had realized that the effect of the lifted tennis ball could be used to propel a boat along in the wind. What, you never heard of the lifted tennis ball effect? Well, then, how about the curve ball in baseball? When a ball is thrown with a spin on it, and the spin axis is perpendicular to the ball's trajectory, a force is created that makes the trajectory change. The ball will move sideways, for example. Or upward. This is called the *Magnus effect*, from the name of the German physicist and chemist Heinrich Gustav Magnus (1802-1870) who first explained the phenomenon. As the ball spins, it drags air along with it by viscosity. The air streamlines are accelerated on the side of the ball where its rotational speed and motion speed are in the same direction, so that their speeds are added together, and they are slowed down on the other side where these two speeds are in opposite directions. Just as in the case of an airplane wing or a sail, the accelerated air creates suction, and, on the other side, the slowing down of the air creates a high pressure area (Q. 143). Together, these two effects create a lift perpendicular to the direction of motion.

The amount of lift generated is proportional to the ball's airspeed and speed of rotation. A rotating cylinder produces the same effect.[8]

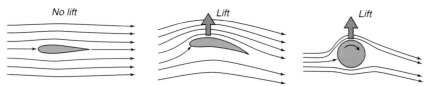

A symmetrical body placed in the axis of a moving fluid (air or water) deforms the flow but does not create a lift (sketch at left). An airplane wing (or a ship's sail) creates an asymmetry in the flow which does create a lift because the upper streamlines are accelerated (center sketch). A rotating cylinder produces the same effect by dragging the air along with it (sketch at right).

Flettner made several trips with the BADEN BADEN to test his idea, first in the North Sea, later in the Atlantic. The motorized rotors were 50 feet tall, almost 10 feet in diameter, and rotated at 700 rpm. The trials were a complete success: the ship could sail up to 25° from the wind, whereas before, with sails, she had not been able to do better than 45°. Still, there was a small inconvenience: the ship could not come about on her own; the auxiliary engine had to be used for that.

---

[8]For a rotating cylinder, lift is given to the first approximation by $L = 4\pi^2 r^2 \rho V N L$, where ρ is the specific weight of the air, r the radius of the cylinder, L its length, V the wind speed, and N the speed of rotation of the cylinder in rotations per second.

The main problem with this system, as with sailing, was that it required wind. It was impossible to compete with steamships for scheduled services, and so the idea was dropped.

But in the 1980's, Jacques Cousteau revived it with the *turbosail*, though a somewhat different principle was applied. In order to avoid the mechanical inconveniences associated with rotating cylinders, Lucien Malavard, who was Cousteau's designer, had the idea of keeping the cylinder stationary and creating the asymmetry in flow by "sucking up" the boundary layer of air surrounding the cylinder. The system was installed on the Cousteau Society's research ship ALCYONE.

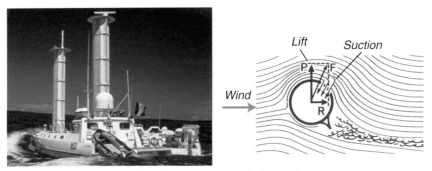

The ALCYONE. *At right, the principle of the turbosail.*   Courtesy of the Cousteau Society.

Another strange boat to cross the Atlantic as recently as 1990 was the BLUE NOVA — this time with airplane-like wings instead of sails. The idea dated from the oil crisis days of the 1970's when economical ways of propelling merchant ships were being sought. Several cargo ships were equipped with rigid sails shaped like airplane wings, hoping to save fuel when the wind was favorable. But the savings turned out to be small, about 10%, and the price of oil having dropped again, that solution ceased to be economically interesting.

At that point, one promoter of the rigid sails concept, an aeronautical engineer named John Walker, turned his attention to yachting and had the BLUE NOVA constructed. This was a large trimaran equipped with wingsails in a bi-plane configuration. She crossed the Atlantic in both directions and, subsequently, four smaller versions equipped with only two wings, the *Zefyr 43* series, were built.

These attempts to develop the wingsail were prompted by the fact that a sail, with it single surface, is not as aerodynamically efficient as a wing. Unlike a sail, an airplane wing has *two* separate working surface, meaning that each surface can be optimized independently to maximize lift.

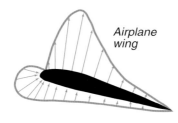
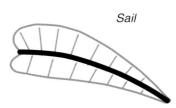

The pressure on the lower surface and suction on the upper surface are higher for a wing than for a sail because the curvature of its two surfaces can be optimized independently.

The *Zefyr* has two wings, the main one and a smaller one acting as a tail. Both are mounted on an arm that can pivot through 360°. The wings are symmetrical, so capable of taking the wind on either side. When the boat is stationary, the tail is aligned with the main wing and they then act together like a large weather-vane. To move in a certain direction, the angle of the tail wingsail is adjusted to obtain the optimal attack angle for the main wingsail. A computer automatically optimizes the attack angle for the desired heading and also reduces

ZEFYR *with* BLUE NOVA *in the background.* *Courtesy of Alison Cooke.*

the angle when the wind increases, to keep the boat from heeling too much. No need to take in a reef anymore... The wings are permanently mounted on the boat, not lowered in port or during storms. When allowed to pivot to align themselves with the wind, they create no more windage than a mast in classic rigging does.

If wings are so advantageous, why don't we see more of them on our boats? Well, there *are* a few inconveniences, including the higher costs inherent in adopting any new system, the fact that, in weak winds, increased efficiency is offset by the increased weight of the wing structure, and the difficulty in removing the wingsail for storage. But the main reason is probably that, as for anything that is new and way off the beaten path, acceptance comes hard, especially in the context of a sport such as sailing where tradition is so important.

There is one other solution that does away with sails completely: an aerial propeller driving an underwater propeller. Equipped this way, a boat could theoretically sail directly upwind. The efficiency is not great, though, because of propeller losses.

137. How are crews distributed on America's Cup boats?

An America's Cup boat, which sails with a crew of 16, is not a place for the faint-hearted or lazy. On the foredeck, two bowmen hoist, control and drop the headsails. Below, the sewerman organizes sails and repacks spinnakers. He is called the sewerman because it is hot, dark and wet inside. Worse yet, the poor guy never gets to see anything of the race. Topside, near the mast, the halyardman and the mastman hoist and drop sails and the spinnaker pole. Then come the grinders and trimmers who work the winches and hydraulic rams. Stationed astern are the navigator who is in charge of data acquisition for wind and boat speed, the strategist who decides on the heading and choice of sails, and the helmsman. A guest might be found at the stern, the architect, for example, or the owner; the guest may not participate in any maneuvers or decisions.

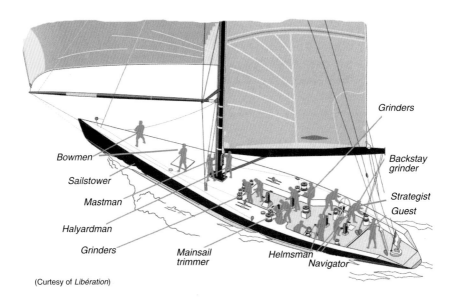

Grinders

Bowmen

Sailstower

Mastman

Halyardman

Grinders

Mainsail trimmer

Helmsman

Navigator

Backstay grinder

Strategist

Guest

(Curtesy of *Libération*)

138. Why is the speed of a boat limited by its length?

Everyone who races knows that the longer the boat, the faster it can go. That is why handicaps are necessary, to give everyone a chance to win. In a good wind, putting up a few extra square yards of sail on a small boat will not make it go any faster. This situation is peculiar to boats. Longer airplanes are not intrinsically faster, nor are longer submarines. The reason why it is true of boats is that they move at the interface between two different environments, water and air.

Just like a submarine, a boat pushes water aside as it moves forward and meets resistance due to friction from the streams of water sliding along its hull. The same thing happens in the air. But a second effect is present at the air-water interface: since air is much lighter than water, it offers essentially no resistance to prevent the water pushed aside by the hull from rising above the surface, too. This results in the bow wave. And a similar situation is found astern: the water pushed down by the hull rebounds up above the surface and creates a second wave, the stern wave.

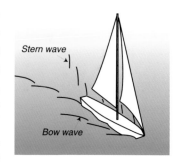

These two wave trains are continually created by the boat's forward motion and thus accompany it at the same speed. But the wavelength of any wave is related to its speed: in deep water, wavelength increases with speed according to the law $V = 1.34\sqrt{\lambda}$, where V is the speed of the wave in knots and λ its wavelength in feet (Q. 14). At slow speeds, the wavelengths of these accompanying waves are short. At 2 knots, for example, the wavelength is about 2 feet, and a bow wave makes several crests along the length of a hull. As speed increases, the number of crests along the hull diminishes. At 4 knots, there will only be 3 crests along the hull of a boat with a 27-ft waterline. And at 7 knots on such a boat, the second crest of the bow wave is located astern and merges with the crest of the stern wave.

Now, when two wave trains meet and merge, the resulting wave is the sum of the two: two coinciding crests form a single, higher crest, while two troughs combine to create a deeper trough. And when a crest meets a trough, the one is subtracted from the other, a phenomenon called *interference* in physics.

When a boat's speed is such that the wavelength of the bow wave equals the length of the waterline, the bow wave combines with the stern wave, the crest of the stern wave is reinforced and the hull sinks into the progressively deeper trough. As speed increases even further, the wavelength of the bow wave stretches out, the boat rears up, the bow riding on the bow wave and the stern down in the trough of the stern wave. The boat is then facing a mountain of water that is very difficult to climb.

| < hull speed | hull speed | > hull speed |

A hull capable of planing can climb its bow wave (Q. 141), but a traditional (displacement) hull cannot. The displacement boat has reached what is called its "hull speed." This speed cannot be exceeded unless the boat is towed by a faster one, but that comes with the risk of catastrophe, since the stresses on the hull are then enormous. The exact value of a hull speed limit depends somewhat on the shape of a hull, but by convention it is considered to be reached when the wavelength of the bow wave is equal to the length of the waterline. With the waterline expressed in feet, the hull speed of a displacement boat is thus given by:

$$V \text{ (in knots)} = 1.34\sqrt{L}$$

To appreciate the difficulty of exceeding hull speed, we need to look at the power required.

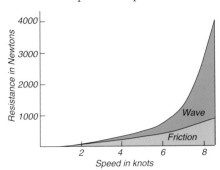

Example of resistance to forward motion for a 13-meter boat in calm water.

Resistance to forward motion due to the bow and stern waves is called the *wave-making resistance*. At slow speeds a boat raises only small waves and the corresponding wave-making resistance is negligible. The overall resistance is mainly due to friction of the water against the hull, and this varies as the square of the speed. In order to go twice as fast, you would need to supply 4 times as much power.

At greater speeds, however, wave-making resistance becomes dominant, and values for this factor increase quickly with speed — roughly as the 6th power of speed, as a matter of fact. To go twice as fast would now require 64 times as much power! For a motorboat, such an increase in power is beyond the reach of the engine. And for a sailboat, extremely strong winds would be required, but then the rigging and sails would not resist.

In practice, a displacement sailboat can only maintain her hull speed limit in ideal conditions: strong winds and protected waters. Under normal cruising conditions, one can count on no better than 70% of the hull speed.

139. Why are catamarans faster than monohulls?

It is tempting to think that if catamarans are so fast, it's because they are somehow exempt from the laws of hydrodynamics that apply to monohulls, or that dividing a hull in two miraculously lessens resistance to forward motion. Naturally, that cannot be the case. The same laws apply in both cases.

As a matter of fact, for a given displacement, a monohull can theoretically perform slightly better than a catamaran. This is because catamarans suffer from parasite effects related to interference between the bow waves of the two hulls.

The problem is that the theoretically optimal monohull, i.e., whose hull shape offers the least resistance for a given displacement, is extremely narrow, with a length to beam ratio of over 10. Such a boat is consequently, in essence, unstable. In practice, then, the monohull, which has to be ballasted and relatively beamy for stability's sake, is far from having an optimal hull. The catamaran, on the other hand, with its intrinsic stability, can have very narrow hulls.

Same displacement

For the same displacement, a monohull could be as fast as a catamaran, provided it had a very narrow hull.

So the catamaran's fundamental advantage is that the normal requirements for stability, beam and ballast, don't apply. Their hulls can be very narrow and have small wetted surfaces, thus minimizing both wave-making resistance and friction.

In a good wind, catamarans can raise their windward hulls out of the water to reduce resistance even further, and then they are unbeatable. The difference is less pronounced in light winds, as then they are resting on two hulls and usually carry less sail than light monohulls do.

140. If an America's Cup boat and a 60-foot Open ocean racer competed against each other, which would win?

The answer to that question is, it depends. Both types of boats are about the same length (60 feet at the waterline) and have about the same sail area (3000 square feet close-hauled, 5000 square feet downwind). But the boats were designed for entirely different programs.

America's Cup boats are built to sail upwind, which represents half of their race course, so they are long and narrow with 80% of their weight in the keel. Beating to windward in relatively calm water, they can do 9-10 knots 30° from the apparent wind as soon as a bit of breeze shows up. No 60-foot Open could get near one in those conditions.

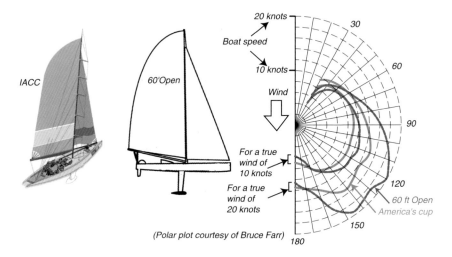

(Polar plot courtesy of Bruce Farr)

Things are different running downwind. America's Cup boats are displacement boats, intrinsically limited by their hull speeds (10–11 knots) no matter how strong the wind may be. There, a 60-foot Open would easily get the upper hand. Their wide beams and canting keels allow them to carry much more sail with appreciably less weight, and they take off planing like racing dinghies. Their speed then essentially depends only on the strength of the wind and how much their rigging and sails can take — they can reach 20 knots and the elegant America's Cup boat would trail there.

141. Exactly what is going on when a sailboat is planing?

For some, the term "planing" might suggest the behavior of a seagull or a glider in the wind. Wouldn't a planing hull be raised up by a lifting force in the manner of an airplane wing?

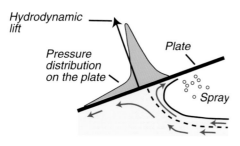

It is not quite the same, however, since a wing is totally immersed in the same fluid (air), whereas a planing boat is moving at the interface between two fluids (water and air). In a wing the lift is due both to suction on the upper side and pressure on the lower side, whereas a planing hull is only supported by the dynamic lift created by the moving water on its lower side. It is the same phenomenon that occurs in water skiing or when car tires are aquaplaning in the rain.

When a boat starts planing, the dynamic lift compensates for part of the boat's weight. The hull lifts itself up, reducing the wetted area and increasing speed. Speed creates more lift and lift, more speed. Eventually, the hull rises above its bow wave and...off it goes!

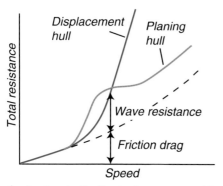

A planing hull eliminates wave-making resistance.

The wave-making resistance is almost eliminated. You can see that because the wake of a planing hull is much reduced, which is why a planing motorboat that passes is much less troublesome than one going at a slower speed.

A boat must be light in order to plane, and its hull must be flat enough to generate dynamic lift. Traditional keeled boats, called "displacement boats" because they continuously displace water as they move, cannot hope to plane. When a displacement boat reaches its hull speed (Q. 138), it rears up behind its own bow wave and cannot go over it, even if provided with more power. This is because its bottom has the wrong shape for generating lift. On the contrary, the wave train it generates tends to push its hull down into the water.

142. What makes a hydrofoil sailboat so fast?

A displacement boat has to overcome two forces that resist forward motion: friction of water against the hull and wave-making resistance (Q. 138). Hydrofoils "kill two birds with one stone" by getting the hull up out of the water, thus eliminating both types of resistance. Planing hulls also rise up, but without getting completely out of the water. In addition, the lift on a planing hull comes only from the pressure on its underside. Hydrofoils raise themselves up on submerged foils which act like sails or airplane wings (Q. 136): their lift comes *both* from pressure on the lower surface and suction on the upper surface.

The hydrofoil elements are set in a V configuration, to counteract drift and ensure stability. Flaps mounted at the rear of the hydrofoils are continually adjusted to counter wave movement. Piloting a hydrofoil is thus a lot like piloting a plane, except that flap adjustments have to be made much more precisely and rapidly.

The pioneer in sailing with hydrofoils is the American Sam Bradfield, who began designing trimarans with hydrofoils in the 1960's and who held the sailing speed record between 1978 and 1982, with 24 knots.[9] For ocean sailing, Hydroptère, the hydrofoil sailing craft conceived by Éric Tabarly and currently in the hands of Alain Thébault, has reached a speed of 40 knots.

These remarkably high speeds are possible because the resistance to forward motion is only due to the drag of the hydrofoils and wind friction against the hull. But speed comes at a price. Where the weight of the hull in a classic boat is supported "free of charge" by the buoy-

Alain Thebault's Hydroptère.

ancy force, a hydrofoil, like an airplane, is held up only because it propels itself along at sufficient speed. In order to rise up on their hydrofoils, these boats require winds above 15 knots and their carrying capacity is very low. Another limitation is that they can only sail in relatively calm seas.

---

[9]Note that Yellow Pages Endeavour, which holds the current sailing speed, is not a hydrofoil; it has three planing hulls.

143. Why is the lee side of a sail under negative pressure?

Most sailors think they more or less know the answer to this question, but actually, the answer is not all that simple. The figure below shows the flow pattern around a sail sailing upwind during a wind tunnel test. To visualize the flow, smoke is injected intermittently, half a second on (colored bands), half a second off.

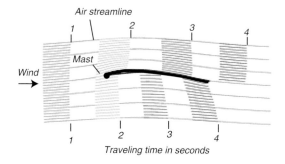

Air flow pattern near a sail. Based on a simulation by John Stewart Denker.

Notice that the air flow is slower on the front (windward) side and faster on the back (lee) side, to the point that the air on the back side arrives at the leech (the trailing edge of the sail) almost a second before the air on the front side does.

The reason for this is that the presence of the sail in the airflow induces a general *circulation* of air around the sail that, superimposed on the free stream, slows it down on the front side and increases its velocity on the back side.

The process starts when the sail is sheeted-in. The streamlines on its windward side are simply diverted, but those on the lee side cannot follow the sail because of its curvature. As a result, they separate from it, creating suction which pulls air from the windward side (sketch *a*, next page) and ends up forming a vortex at the trailing edge of the sail. This vortex, called the "starting vortex," is swept downstream, pulling the surrounding air into motion by viscosity, just as a gear would, and triggering a large circulating flow of air around the sail (sketch *b*). The starting vortex is shed once this circulation is established, but the circulatory flow persists.

Circulation was originally a mathematical concept for calculating the lift of an airplane wing,[10] but the phenomenon is a real one: air does actually circulate around a sail.

---

[10]From the beautiful Kutta-Joukowski theorem which dates from 1910, the lift per unit length of a wing is equal to the product of air speed, air density and circulation.

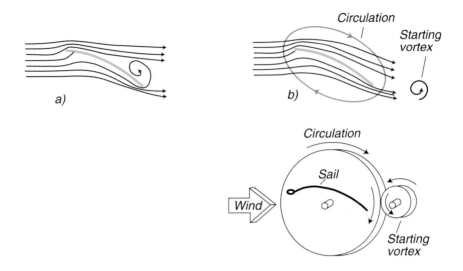

You can convince yourself that the air does indeed circulate by doing the little "kitchen" experiment (actually, a bathroom experiment) proposed by the aerodynamicist Arvel Gentry.[11]

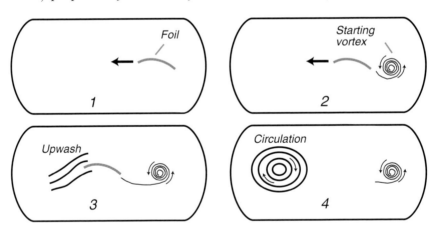

Put about 4 inches of water into a bathtub and allow it to settle. Next, gently sprinkle fine sawdust or talcum powder on the water's surface so as to better visualize the flow. Then, using a piece of waxed cardboard (cut from a milk carton, for example), make a small section of a foil about 6 inches wide and camber it into the shape of a sail. Gently dip the foil halfway into the water at one end of the bathtub and move it slowly towards the other end (sketch 1). Notice that a vortex appears at the trailing edge and remains in the same spot as

[11]For a detailed description of this experiment, see *The art and science of sails* by Tom Whidden and Michael Levitt [83]).

the foil moves away. This is the "starting vortex" (sketch 2). As the foil moves forward, notice that the water in front of it seems to be "aware" that the foil is coming and curves "upwards" in order to pass over it (sketch 3). Aerodynamicists call this *upwash*. Before the foil reaches the far end of the bathtub, lift it out of the water. At that point, there appears in its place a slowly rotating flow turning in a direction opposite to that of the starting vortex (sketch 4). This is the "circulation" flow. The upwash is no longer mysterious: the premonitory motion of the water was simply due to this circulation flow.

The circulation flow also explains why flow is faster on the back of a sail than on the front: on the back side, the rotating motion adds itself to the free flow of air, while subtracting itself on the front side. Remember now that, in any flow, a locally higher velocity increases the pressure while a locally lower one creates suction. This is the so-called Venturi effect.[12]

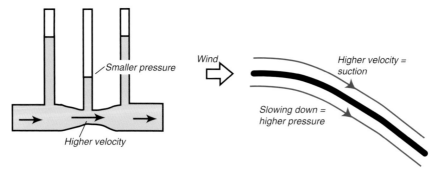

A higher flow velocity in a constricted pipe decreases the pressure — this is the Venturi effect (left). Similarly, the increased speed on the back of a sail creates suction, and the slower speed on the front creates a supplemental pressure (right). The resultant of this is the "lift" (propulsive) force on the sail.

We finally have the answer to our question: the presence of a sail in a flow of air creates a "circulation" of air around the sail. On the lee-side, this increases the velocity of the air's motion, hence creating suction. On the windward side, this decreases the air's velocity and increases the pressure on the sail.

---

[12]This stems from Bernoulli's law which states that the energy total is conserved along a streamline. Since the total energy is the sum of the kinetic energy, which is a function of the square of the velocity and of the potential energy (pressure), if the velocity increases, the pressure diminishes.

144. What is the slot effect?

There is no doubt about it: with the mainsail alone, a sailboat is rather sluggish. Same thing with the jib or genoa alone. But put the two together and the boat wakes up. It really does seem that jib + mainsail is a winning combination.

Wrong ! Acceleration

According to popular belief, this little miracle would be due to the "slot effect," which is usually explained as follows: the wind rushes into the slot between the jib and the mainsail, its speed increases due to the funnel (or Venturi) effect, and this, in turn, increases the suction on the back of the main and hence its lift. This explanation, which is repeated far too often, is simply not true. As the aerodynamics engineer Arvel Gentry showed in the 1970's, wind does not accelerate inside the slot and the mainsail lift is not improved [24, 83]. On the contrary, it is the jib that benefits from the presence of the mainsail.

This becomes clear if one remembers that a sail in the wind generates a *circulation* pattern of air around itself (Q. 143). The mainsail and the jib generate two circulation patterns, both turning in

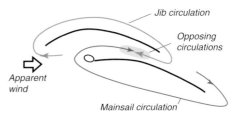

the same direction and thus opposing each other in the slot between the main and the jib. The airflow in the slot is not accelerated, but slowed down. Some of the air that we expect would flow into the slot is in fact diverted by the combined circulation and passes on the lee side of the jib, increasing its efficiency.

This is the global view, but it is instructive to examine in detail what happens to the streamlines around the two sails.[13]

Let us first have the mainsail up by itself. Its presence is actually felt well upstream and the approaching streamlines change direction in preparation for passing the sail. The streamline marked S stops at the mast (the "stagnation" point) and separates the air that passes on either side of the sail. We notice that the streamlines on the lee side are closer together than on the windward side. According to Bernoulli's law, this produces negative pressure on the lee side of the

---

[13]This is a brief summary of the analysis done by Arvel Gentry in an excellent series of articles published in Sail Magazine in the 1970s and which have been collected in the book *The Best of Sail Trim* [24].

sail which contributes to the lift. The point marked H is the location of the headstay onto which the jib will be raised. The distance between the streamline passing by H and the stagnation streamline S is a measure of the amount of air that goes through the "slot" when the jib is not there.

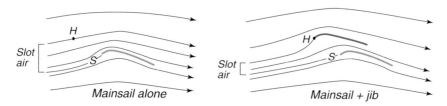

Now, let us raise the jib. We notice that the stagnation streamline for the mainsail has moved slightly to leeward, and that the stagnation streamline for the jib has moved significantly windward of the old streamline H. This indicates that the amount of air going through the slot is *less* with the jib than without it, the exact opposite of the folkloric venturi "explanation." Where did the missing air go? It went on the lee side of the jib, forcing the streamlines closer together (because more air had to go through), increasing suction and therefore lift. We also notice that the streamlines arriving on the windward side of the jib are deflected favorably by the presence of the mainsail, and thus that the boat can point closer to the wind without the jib luffing.

In summary, the only beneficiary of the jib/mainsail association is the jib. Its efficiency increases by about 50%, and this at the expense of the mainsail.

The favorable position of the jib compared to that of the mainsail is similar to the "favorable leeward position" used in sailboat races to stay ahead of a pursuing boat. Just like a jib, boat A receives more air and in a more favorable direction than boat B and can thus block any attempt to overtake.

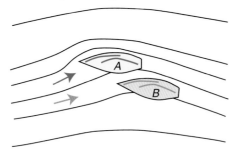

With respect to the mainsail, the jib is in the same "favorable position" as the leeward boat A is to boat B.

145. How efficient is a spinnaker?

If an old Cape-Horner could see one of our modern boats display-ing a spinnaker, his eyebrows would surely shoot up. "What's that gaudy balloon thing? Go get a good square sail!"

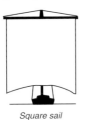
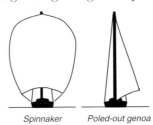

Square sail　　　　Spinnaker　　Poled-out genoa

For running down-wind, it's true, noth-ing beats a square sail on a yard high on the mast; it presents the largest surface area to the wind. But who to-day is willing to climb the mast to set one? So, things being as they are, the spinnaker is a good second choice. It requires no yard and, although it presents a smaller surface than a square sail to the wind, it beats a poled-out genoa.

The upper part of a spinnaker[14] is shaped like part of a sphere so that air pressure in-side creates a lifting force to compensate for the weight of the sail and sheets. The spherical form is advantageous, too, as it has a better drag coef-ficient than a plane surface does, hence a greater

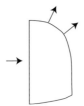

propulsive force (the propulsive force being equal to the cross-section area multiplied by the drag coefficient).

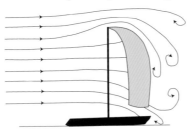

Although indispensable for downwind conditions, the spinnaker is not an efficient sail. Per square yard, mainsails and jibs are more efficient (running upwind) than spinnakers (running downwind). In fact, in a spinnaker, only the windward side helps propel the boat by its drag; wind turbulence that forms behind it prevents the back side of the sail from contributing. In the case of a fore-and-aft sail (main or jib), on the other hand, both sides contribute: the windward side by higher pressure, the back side by suction (Q. 143).

The difference is the same as that between a propeller and a paddle-wheel. A propeller works like an airplane wing or a fore-and-aft sail: the front side of the blade works by suction, the back side by pressure. In a paddle-wheel, only the sides of the paddles that push against the water do any work. For the same amount of energy

---

[14]In 1870, an English boat, the SPHINX, is reputed to have been the first vessel to use this type of sail, and a deformation of the "Sphinx's acre" — the "acre" being a reference to the enormous size of the new sail — is said to have given us the term *spinnaker*.

expended, propellers are much more efficient. The year 1845 saw a brilliant demonstration of this truth. At that time, since both the paddle-wheel and the propeller had passionate supporters, the British Admiralty decided to settle the matter by organizing a contest. It equipped two identical ships, one with a propeller and the other with a paddle-wheel, and made them compete in various tests. The RATTLER, with the propeller, won systematically over the ALECTO, with the paddle-wheel. The final combat was a tug-of-war: a line was tied to the stern of each ship and their engines were turned on full. The RATTLER was the undeniable winner: she dragged the gasping ALECTO at almost 3 knots...in a head wind, to boot.

Tug of war between the RATTLER, *with propeller (left) and the* ALECTO, *with paddle-wheel (right).*

The poor efficiency of the spinnaker coupled with the reduction of apparent wind often makes it preferable to replace the traditional spinnaker with an asymmetrical one and to tack downwind. The gain in speed largely compensates for the greater distance traveled: the VMG (*velocity made good*) is better.

146. Why do sails luff in the wind?

When the wind is steady, a weather vane, with its flat, rigid surface presented to the wind, oscillates slowly and very little, whereas flexible surfaces, like flags and sails, flap rapidly. Why the difference? This simple question has had aerodynamics researchers stumped for over 100 years.

One might imagine that the mast supporting the flag could be the cause of the flapping. A mast perturbs the air flow, and vortexes peel off from it on alternating sides. These are called *von Karman vortexes*. It is they that cause telephone wires and guy wires (shrouds, stays) to hum in the wind: as they detach themselves periodically from the cable, the vortexes pull it from side to side, making it resonate like the string of a musical instrument.

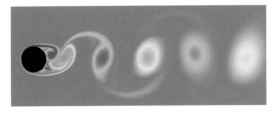

Alternating shedding of vortexes behind a mast.

But recent laboratory experiments have shown that this theory fails to explain what happens with a piece of cloth [88]. First of all, the vortexes do not start at the leading edge of a sail or flag, but at its trailing edge. Moreover, unlike what happens with vortexes created by masts or cables, the ones trailing off behind a flapping piece of cloth do not rotate in opposite directions, alternating every other one. Instead, several vortexes rotate in the same direction, and then, after a half-cycle of the beat, several others form in the reverse direction for the second half-cycle.

Laboratory simulation of a flag flapping. The flow is from left to right. Note that the vortexes detach from the trailing edge of the flag. Courtesy of Jun Zhang, New York University.

So how to explain luffing? For the moment, there are no clear answers. Experiments are hard to do, and the mathematical equations describing the interaction between wind and a flexible body are extremely complex. Still, it does now seem that luffing is related to the inertia of the cloth and its elasticity when deforming under the action of an airstream.

This phenomenon has some elements in common with the way fish swim. There, too, there is interaction between a fluid and a flexible body. The swiftest fish swim in such a way that their natural oscillations work in harmony with the disturbances they create in the water as they move through it (Q. 52).

147. How do the different sail rigs (sloop, cutter, ketch, schooner) compare in efficiency?

Thirty to forty years ago, the comparative merits of various sail rigs were debated interminably. The sloop rig had the virtue of simplicity. The cutter, with its two headsails, brought flexibility in sail reduction. The ketch and schooner rigs increased this flexibility even more, and their proportionally smaller sails were easier to handle, besides making it possible to add sail area between the masts when running downwind. In general, the bigger the boat, the more the sail plan was subdivided, and, likewise, for a given boat size, the smaller the crew, the more the sails were subdivided.

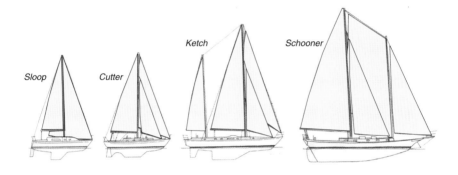

All that changed with the arrival of headsail furlers, mainsail furlers, reduction winches, and other niceties. Very tall masts became possible, and maneuvering enormous sails no longer posed a problem for crews of men (and women) with ordinary muscle-power. No matter how big the boat, the sloop rig has now become the standard. Its aerodynamic efficiency is better, one tall mast is less expensive than two, and the disappearance of overhangs (Q. 134) means that an extra mast is difficult to accommodate, in any case.

In practice, however, how do these different types of rigs compare?

The differences are small for running downwind, but they are noticeable when sailing upwind. The main reason for this is that only the leading sail is in a clean wind flow. The second sail has to be sheeted in more because of the slot effect (Q. 144), and the third sail even more so. About 5° are lost with respect to the wind for each successive sail, resulting in two negative consequences: the propulsive component drops and the heeling component increases. It is generally felt that, compared to the sloop, the aerodynamic efficiency of the ketch is 96% and that of the schooner 92%. This is so true that, close hauled, a ketch is often faster without the mizzen.

To avoid this sail interference problem, masts have to be set far enough apart so that the wind in each sail is undisturbed by the wind in the preceding one. Jean-Yves Terlain's three-master VENDREDI 13 managed this, and so did the ketches with well-spaced masts (more yawl-like) designed by Bruce Farr, which won the 1990 Whitbread.

Nowadays, the only remaining advantage of subdividing the sail plan is that it lowers the center of thrust. The boat heels less and can carry more sail in a strong wind.

148. Why is the modern sailing rig called a "Marconi rig"?

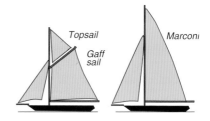

The Marconi rig, the one universally adopted today, represents an evolution of the gaff rig to which a topsail had been added. It made its appearance in yachting around 1920 with the aim of making maneuvers easier and eliminating the dangerous gaff that swept the deck when the sail was lowered. The Marconi rig has the added advantage of being more efficient when sailing upwind because of its higher aspect ratio.

The rig's name comes from its resemblance to the radio antennas used early in the 20th century by the physicist (and Nobel prize winner) Guglielmo Marconi (1874–1937), one of the pioneers of radio. Marconi succeeded in making the first successful transatlantic radio transmission in 1901, and, as of 1910, most steamships were equipped with radio sets based on his principles.

In 1919, Marconi bought the steam-powered yacht ELETTRA, which he transformed into a floating laboratory to study the transmission of short waves, and that he used for worldwide communication. He is thus the father of the short wave radio communication system used at sea today.

Marconi's yacht ELETTRA, *with her two masts supporting the antenna.*

The Marconi rig was not a new idea, however. Since 1800, this type of rig had been common on English schooners sailing to the

Caribbean and Bermuda, whence the name *Bermuda rig* that is sometimes also used for it.

Bermuda schooner in 1834 (from a watercolor by J. Lynn [12]).

149. What is the advantage of a raked mast?

In modern boats, the *rake*, or amount a mast is angled towards the stern, is between 1 and 2 degrees. This is done for two reasons: first and most importantly, the sail hangs better under its own weight in light winds. Then, when lowered, it also falls more naturally.

In a ketch or schooner the aft mast is angled one degree more than the forward mast. The reason there is purely aesthetic: because of an optical illusion, two exactly parallel masts appear to converge.

American schooners of the 19th century had very raked masts, ten degrees for the famous AMERICA. That was for a different reason: since the decks of those schooners were low on the water and the mainsail booms were extremely long, the masts were raked to keep the boom from plunging into the water when the boat heeled.

150. Why are sail panels laid perpendicular to the leech?

As any seamstress will tell you, cloth stretches very little in the direction of the woof or weft but is easily stretched out of shape when pulled along the diagonal, or bias. Try it yourself with your napkin.

A piece of cloth stretches much less in the direction of the woof or weft than along the diagonal.

A mainsail is stretched out of shape by pressure from the wind, its center being "pushed in" the most. The figure at right shows the main lines of stress in a typical mainsail. Even though it is primarily supported at its three apexes (head, clew and tack), the boom and mast hold it firmly, so that the strongest stress lines run alongside the leech. To keep the cloth from deforming even further due to the bias effect, the canvas strips are placed in such a way that the threads run perpendicular to the leech.

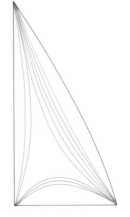

They can be set either parallel to or perpendicular to the leech. In the old days, they were set parallel to the leech in order to minimize the risk of the seams splitting under tension. But early in the 20th century it was discovered that, in cloth, the fill threads (those running side to side) stretch less than the warp threads (those running lengthwise). So to minimize the stretching of the sail under wind pressure, the practice of setting the crosswise threads perpendicular to the leech began, and this practice continues to be used in cloth sails today.

This does not apply to Mylar or Kevlar sails, as they essentially deform to the same extent in the direction of warp and fill. That being the case, the current preference is to set the panels parallel to the direction of the greatest strain, i.e., parallel to the leech, as

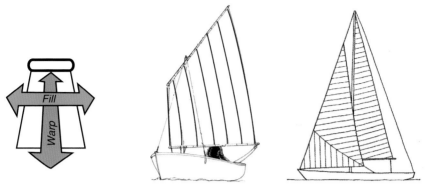

At left, a bolt of sailcloth showing the directions of warp and fill. At center, a typical, late 19th century sail with the panels parallel to the leech, and, at right, a modern sail with panels perpendicular to the leech.

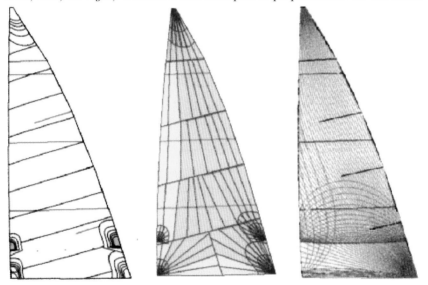

From left to right, cloth sail, mylar sail, and "3DL" sail.   Courtesy of North Sails.

that allows the weakest elements, the seams, to run along the lines of greatest stress, not across them.

For racing, where one wants the sails to deform as little as possible in the wind, the sail is made up of a large number of narrow panels which are oriented in the direction of the lines of maximum *local* stress. The ultimate solution consists of completely eliminating the bias effect by replacing the panels of woven canvas with resistant fibers covered with film on each side. This makes it possible to give the fibers ideal orientations to minimize deformation, and the covering film simply distributes the strain and keeps the air from passing through. This is the "3DL" process perfected by North Sails. Compared with traditional sails made of cloth strips, 3DL sails are all of a piece, 30% lighter, and stretch less. But the process is complicated, requiring molds and special ovens, so is normally reserved for high-level races.

151. Why are marine propellers so different from airplane propellers?

Marine and airplane propellers are indeed very different. Plane props have long, thin blades, whereas marine propellers have broad, stubby blades.

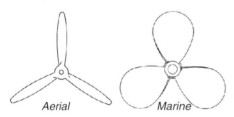

Ideally, marine propellers should be more like their aerial cousins, as theirs is the more efficient shape. That is impossible, however, because of the limits imposed by a boat's draft: the diameter of a marine propeller has to be small enough to keep the blades from descending deeper than the bottom of the keel and from coming so close to the surface that it churns in the air. Now, reducing diameter also reduces thrust, and to compensate for that, the blade area has to be larger.

152. What causes "propeller walk"?

It can hardly have escaped your notice that, when you accelerate your engine, the boat drifts sideways. The effect is especially noticeable when you are just starting out. A number of subtle factors can contribute to this phenomenon, but it is mostly due to the propeller shaft being angled downward, which is usually the case with inboard engines.

A propeller with its plane of rotation perpendicular to the direction of motion produces a balanced thrust with no tendency to push the boat sideways. If the propeller shaft is tilted downward, however, blades on one side will have a larger pitch than on the other side, leading to an imbalanced thrust. Let us take the common case of a right-handed propeller, meaning that it turns clockwise when in forward gear. The descending (starboard) blade will have a larger pitch angle than the ascending (port) blade. Larger pitch means larger an-

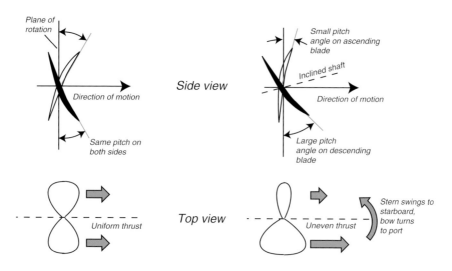

Propeller walk is mainly due to tilted shafts. At left, a propeller with an horizontal shaft produces a balance thrust. At right, a propeller with a tilted shaft produces an uneven thrust because, for a right-hand propeller, the ascending blade has a smaller pitch than the descending one.

gle of attack and therefore larger thrust. As a result, the starboard side of the propeller produces more thrust than the port side and this induces a yaw force to port.

Pilots of small planes are very familiar with this effect: during ascent, with the engine on full power and the plane's nose up, they have to push firmly on the rudder to keep going straight. On a boat motoring at constant speed, the helmsman unconsciously corrects for the propeller's imbalance. But when starting out, the rudder has little controlling power since water speed is almost nil, and you really feel the boat swinging sideways when you rev up the engine.

On a plane, the effect is called the *p-factor*. On a boat it is called *propeller walk* because the sideways thrust is in the same direction that it would be if the propeller were hitting the bottom and "walking" on sand or mud. Propeller walk can come in handy to assist in kicking the stern away from the dock. It can also be used to turn "on a dime" in a crowded harbor, by accelerating alternately in forward and reverse. In reverse, propeller walk swings the stern to port (for a right hand propeller). Switching to forward will cancel the backward motion and help continue turning the boat to starboard provided that the rudder is turned to starboard to benefit from propeller wash against it.

On boats with two engines and two propellers, the propellers turn in opposite directions to neutralize the propeller walk effect.

153. What makes a winch so effective in pulling a sail?

Winches are everywhere on modern sailboats, on halyards, jib sheets, reef lines, etc. Thanks to the winch, there is no longer any need for the large crew of big, burly musclemen that used to be indispensable. Like block and tackle, winches procure a mechanical advantage, but they have a big plus: block and tackle gears give a mechanical advantage but at the cost of hauling in rope in direct proportion to the reduction of effort; a winch with a crank handle provides the same mechanical advantage without the need to haul in yards and yards of rope. We owe this small miracle to *friction*.

Friction most often works against us. It's what slows down our hulls in the water, what creates drag in our sails, what dissipates the energy of our engines uselessly. But for once, in the winch, friction works for us...exponentially, even!

If you pull with the force T_0 on the tail of a rope wrapped on a winch, the tension in the rope on the loaded side (e.g. by a sail) can be as high as

$$T_1 = T_0 e^{f\alpha}$$

before the rope begins to slip, with α being the angle that the rope makes around the drum and f being the coefficient of friction between the rope and the drum.[15]

The ratio of forces between the rope under tension (sail side) and the tail end (crew side) goes up considerably with the number of turns around the winch, as can be seen in the table below. It is not for nothing that novices are advised to "give it three turns!" Note, too, the serious loss of friction (and hence mechanical advantage) when a rope is worn or wet.

| | | | Force ratio | |
Rope	f	1 turn	2 turns	3 turns
new and dry	0.3	6.6	43	284
worn and dry	0.2	3.5	12	43
worn and wet	0.1	1.9	3.5	6.6

---

[15]This is easy to demonstrate. If the rope under tension T lies on the drum over an angle $d\alpha$, the force it exercises radially on the drum is $T d\alpha$ (by vector addition). The reduction of tension dT due to friction is thus $fT d\alpha$. By integrating over the angle α, we get, $\alpha = 1/f \text{Log}(T_1 - T_0)$, whence $T_1 = T_0 e^{f\alpha}$. Note that the ratio of the two tensions (T_1/T_0) is independent of the diameter of the winch's drum.

When cranking a winch, the mechanical advantage depends on the ratio of the length of the handle to the radius of the winch drum (b/a), typically about 4 or 5. When the winch has two speeds, the gain is even greater. One would need tackle blocks with 3 pulleys to arrive at the same force ratio, and it would be impossible to haul in rapidly.

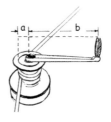

154. What is the difference between a sea-anchor and a drogue?

Sea anchors and drogues are both devices that are thrown into the water during storms at sea to slow a boat down. A sea anchor is a canvas cone with an open end, while a drogue can be a reduced model of a sea anchor or even simply long ropes dragging behind. But the two systems work on different principles and have different "directions for use."

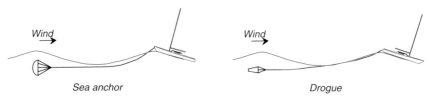

Sea anchor   Drogue

A sea anchor is deployed forward, the drogue aft.

A sea anchor is made fast to the bow to keep the boat facing into the waves, thus presenting its best defenses without risk of turning broadside into the weather. It must be strong enough to practically *stop* the boat (with a maximum reverse speed of about one knot) so as to avoid damaging the rudder which is then working backwards.

A drogue, on the other hand, is thrown out aft with the intention of simply *slowing down* the boat, not bringing it to a halt. The aim is to prevent the boat from surfing the crests of the waves and keep it under control so that, once again, it does not turn broadside to the weather and risk capsizing. It is important not to slow the boat down too much, however: a speed of at least 3 knots has to be maintained if the rudder is to remain useful. Another condition is that the drogue not be made fast too far behind the center of lateral resistance, at least not aft of the rudder, or the rudder will become useless. Some long distance navigators, Moitessier in particular, have been disappointed by their drogue experience, but that is because they used them without respecting these conditions [30].

Ideally, the sea anchor (or drogue) should be set at a distance equal to the average wavelength of the waves so that the orbital movements of the water under the boat and around the sea anchor are in harmony (Q.14). As indicated in the table below, the orbital speed of the water is very large in heavy wind conditions. In force 8, for example, if the sea anchor is in the wave trough while the boat is up on the crest, the pull of the towline would be as strong as if the boat were being towed at a speed of 10 knots, causing serious jolting and jarring. If the length of the towline is equal to the wavelength, however, this effect is minimized, and no additional tension will be created.

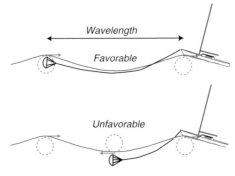

Beaufort scale	7	8	9	10	11
Orbital speed (knots)	4.2	5.2	6.3	7.4	8.2

155. Why do you have to give rope a twist when you coil it?

When coiling a rope, if each coil is not given a twist, it will not lie flat and docile for you, but turn into an unmanageable snarl.

To understand why this happens, it is helpful to take a piece of rope and mark a series of dots along a length of it as it lies flat. It is then easy to see how it twists in your hands as you manipulate it.

As you take up the free end to make a coil, it turns out that to keep from giving the rope an unwanted twist, your hand would have to swivel completely around. Since human anatomy does not allow this, one usually just passes the rope to the hand holding the previously made coils. In doing so, the rope picks up a twist, the coil turn itself into a figure eight, and there is every likelihood that it will become snarled with preceding coils.

To avoid this problem, the rope being hauled in must be twisted 360 degrees between the fingers before being passed to the other hand. Alas, even your best efforts to judge the amount of twist to give each coil are often not good enough. The coils are rarely perfect

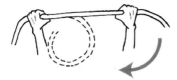

To coil a rope without twisting it, the hand holding the free end would have to pivot around, making a complete turn, which is anatomically impossible.

and may still tend to snarl when undone. This is why ferrymen don't usually coil a rope that they may soon have to send back. As they haul in a mooring line, they let it fall on deck and pile up naturally. It may not look neat, but a rope treated this way can be sent back without risk of snarling as it pays itself out.

156. Why don't the strands in a rope untwist?

Rope is no more than twisted fiber strands. What keeps them from coming untwisted?

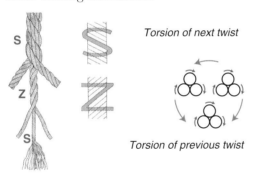

Torsion of next twist

Torsion of previous twist

A twisted rope is made in three stages. Fibers are first spun to form yarns, which are then twisted together to form strands, and finally the strands are twisted together to form the rope. The secret of rope making is that the twisting is done each time in the direction opposite to that of the previous twist. If, for example, yarns are twisted clockwise (called an *S*-twist), the strands will be twisted counterclockwise (*Z*-twist) and the rope twisted with an *S*-twist. The torsion of the rope is balanced by the internal torsion of the strands which is itself balanced by the internal torsion of the yarns, and, presto!, the rope remains twisted. There is one condition: the individual strands must not be allowed to untwist at the two extremities of the rope. This is ensured by ligature or, in the case of synthetic materials, by burning the ends to fuse them.

Navigation

157. What do all those radiating lines on ancient nautical charts mean?

Prior to the 12th century, navigation in Europe was largely coastal and charts were unnecessary; to sail from one port to another, seafarers simply followed the succession of recorded landforms and landmarks that showed the way. Then, with the appearance of the magnetic compass,[1] it became possible to sail directly from one port to another and charts became necessary.

A nautical chart dated 1325.

The first charts appeared in Italy in the late 13th century. Called *portolanos* (from the Italian, a list of ports), they were the pilot books and sailing directions of their day. Those first charts were based strictly on empirical knowledge and showed no grid of meridians and parallels. Instead, the 32 directions of the compass were represented, not just once but in several different places. It is thought that a navigator would place a straight edge ruler on the line connecting his port of departure with his port of destination, then decide which compass bearing to follow by finding the best match between the direction determined by his ruler and a compass direction marked on the chart.

---

[1]Tradition has it that the compass was imported from China, but there is no tangible proof for its use there prior to the 12th century, and it is probable that it was invented independently in several places.

158. When did modern maps showing latitude and longitude first appear?

In ancient times the Egyptians and Babylonians drew maps, but used them primarily to resolve certain practical problems like establishing borders. The first geographers were the Greeks. A lack of arable land in their country forced them into exploration by sea, the development of trade, and the establishment of colonies. Thanks to their knowledge of astronomy and their philosophical pursuits, the Greeks became the first to show an interest in making maps of the world.

We owe our understanding that the world is round to Pythagoras (6th century B.C.). His contemporaries had already noticed that, when a ship sailed to the horizon and its hull had disappeared from view, the top of its mast remained visible. They had also observed that, during eclipses of the Moon, the shadow of the Earth on the Moon is round. To Euclid (3rd century B.C.) we owe the science of *geometry*, literally, the "science of the measurement of Earth." To Hipparchus (2nd century B.C.) we owe the first star catalog using celestial latitude and longitude, which eventually led to the idea of mapping the Earth, too, using latitude and longitude.[2] And to Ptolemy (2nd century A.D.) we owe the first relatively correct representation of the world.

A copy (engraved on wood in 1545) of Ptolemy's map of the world, the first to show latitude and longitude. Longitude is counted from the meridian that passes through the Canary Islands.

---

[2]These terms come from Latin: *latus* (width), and *longus* (length), respectively.

Ptolemy (Claudius Ptolemaeus, 2nd century A.D.), a Greek astronomer, mathematician and geographer from Alexandria, was incontestably the greatest geographer of the ancient world. He knew that the Earth was round, of course, and after spending many years in the great library of Alexandria studying maps and travel accounts, he published his famous *Guide to Geography* which gives the latitudes and longitudes of over 8000 places on Earth. His work summarized all the knowledge of the ancient world. Unfortunately, it was forgotten in the Middle Ages, during which time most people held the belief that the Earth was flat.

The first marine charts date from the 13th century (Q. 157), but those were highly empirical. It was only after the rediscovery of Ptolemy's description of the world, early in the 15th century, that objective descriptions using meridians and parallels came back into use.

Now if it is easy enough to plot latitude by sighting the North Star (polestar) or the Sun at its highest elevation, plotting longitude calls for the use of a very precise instrument for measuring time, and this the Greeks did not have. They were reduced to using dead reckoning to estimate the distances covered east-west, and the result was that Ptolemy made a significant error when he drew up his map. He estimated that Europe and Asia covered half the globe, 180° of longitude, whereas it is actually 130°. That error had the consequences we are all aware of when Christopher Columbus set bravely off for China thirteen centuries later...

159. How did seafarers manage to navigate before the invention of the marine chronometer?

Before the invention of modern navigational aids based on radio waves, such as Decca, Loran, and GPS, a navigator could only determine his position by using one of the two following methods:

- Starting from a point for which the exact position was known, careful track was kept at all times of the ship's direction and speed, plus any currents encountered. This is the method known as *dead reckoning* (and which is now used in inertial guidance systems on modern planes and rockets).
- The latitude and longitude of one's position are determined by astronomical means.

The first method became reasonably precise with the arrival of the magnetic compass in the 12th century: the navigator carefully logged any changes in heading and speed, and, at the end of the day, converted his observations into distance and heading and noted them on his sea chart. Unfortunately, errors can accumulate quickly at sea, especially when you do not know your exact heading, which was the case in those early times. For one thing, compasses then were only precise to within about 10° (the compass card, which we now graduate in degrees, was in those days divided into 32 "points" 11° 15′ apart). And for another thing, it was a rare navigator indeed who took into account the difference between the geographical and the magnetic poles (magnetic declination), and that factor, too, adds an uncertainty of about 10°.

The second method, astronomical observation, allows positions to be determined very precisely but requires the accurate measurement of time in the determination of longitude, and this was not possible before the 18th century (Q. 162). Until then, astronomical observations could only supply latitude.

Latitude determination with an astrolabe

So until the 18th century, navigators on long voyages relied on a mix of these two methods, determining latitude astronomically and estimating longitude by dead reckoning. They used a technique called *latitude sailing*. On leaving port, they would sail along the coast to the latitude of their destination, then head out to sea and sail on that parallel to their target port. This more or less guaranteed that they would reach their destination no matter how big their dead reckoning error. Thus, they would really only use dead reckoning to give them an idea of the distance they still had to cover.

This was the method the Arabs used in the Indian Ocean, where the monsoon winds blow from the east or the west for months on end. All they had to do to reach their desired port was simply sail to the proper latitude and remain on it, confirming their position on it by taking daily sightings with their *kamal* (Q. 163). They were then sure of reaching their destination, carried in with the wind at their backs. And for the return journey, they simply waited for the winds to reverse, then followed the same steps.

This was also the strategy that Christopher Columbus used when he set sail for China, first going down to the Canary Islands which, according to Ptolemy's map, were at the same latitude as China, then staying on that latitude for his Atlantic crossing. For the return, he used the same technique, moving progressively north to the latitude of Cape St. Vincent, at the southern tip of Portugal, then following this parallel back across, stopping only in the Azores, a Portuguese colony, after encountering a storm. It was a lucky thing he took that route, too. If he had decided to take his original route back, he would have been beating into the wind all the way, whereas by taking the northerly route, he had the westerlies working for him.

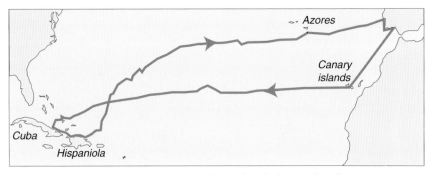

Christopher Columbus's "latitude sailing" during his first voyage.

The Spanish fleets that soon began shuttling back and forth between Spain and the Americas continued to use this same technique for two centuries. And the English, when they started going to the Caribbean, had a saying: "Sail south until the butter melts, then head west."

Latitude was determined by measuring the altitude of the Sun above the horizon at noon and correcting for the day of the year (solar declination), or by measuring the altitude of the North Star and applying the correction coefficient for the discrepancy between the North Star and the celestial pole.[3] With the instruments of that time, the astrolabe and the cross staff, measurements were precise to within a quarter or a half of a degree, about 15 to 30 nautical miles.

---

[3]In the 16th century this correction was not negligible: 3.5°; the discrepancy is less than one degree today, since the position of the celestial pole has changed due to the precession of the equinoxes.

160. How did the ancient Polynesians, in their search for new islands, dare undertake such incredible sea voyages across the largest ocean on Earth?

When European explorers made first contact with the Polynesians, they were still in the Stone Age, without metal, the wheel or a writing system, and their outrigger canoes and instruments of navigation seemed exceedingly "primitive." How could they have possibly managed to colonize that gigantic Pacific triangle 4000 miles on a side that we call Polynesia,[4] between New Zealand, Hawaii and Easter Island? And where did they come from, anyway?

In the mid twentieth century, Thor Heyerdahl launched the popular theory of colonization by people from South America. Considering that the winds and currents in those latitudes move from east to west, that seemed irrefutable logic. In order to prove his theory, Heyerdahl set out from Peru in 1947 on a balsa wood raft, the famous KON TIKI, and allowed himself to be carried by the winds and currents over a distance of 4300 nautical miles, arriving in Papeete, Tahiti, 101 days later (an average speed of 42 miles a day).

But logical-sounding or not, this theory doesn't in the least agree with the archaeological, linguistic, and genetic data that has been collected; the experts now generally agree that the Polynesians originated in southeast Asia (some think Japan), and peopled their Pacific island triangle by "hopscotching" their way through Melanesia, Indonesia, and the Philippines [38].

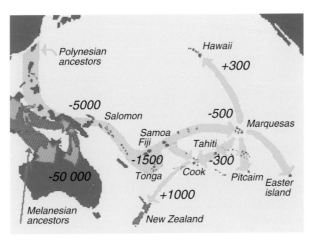

Probable Polynesian migration routes.

---

[4]Polynesia means "many islands."

They are thought to have begun their eastward migration because of population pressure or war, and, after Melanesia and Fiji, moved on to Samoa and Tonga, the true cradle of the Polynesian culture, in about the year 1500 A.D. From there they went to the Marquesas Islands, then to Tahiti, Hawaii, Easter Island, and finally New Zealand.

Still tending to underestimate the navigational prowess of these people of the Pacific and their ability to sail upwind and against currents, archaeologists long held the view that the islands were colonized "by accident," that thanks to storms or navigational mistakes, a canoeful of Polynesians would land by chance on an uninhabited island and settle there.

This kind of thinking managed to ignore the fact that the Polynesians had brought their families, crops, and animals (pigs and dogs) along, too. No, these were certainly not "accidental migrations." How in the world could anyone think that islands as distant and isolated as Hawaii and Easter Island could be discovered purely by accident!

A Tahitian "pahi" (large, double-hulled canoe).
Courtesy of Herb Kane, HawaiianEyes.com.

The current theory is that the voyages of migration were undertaken deliberately [35]. After spending thousands of years in Melanesia, where the islands are so close and numerous that one can practically navigate by sight, the Polynesians developed their big, seafaring canoes and sharpened their navigational talents, then were ready to undertake their great voyages on the high seas.

But what about the trade winds? Wouldn't they have to be sailing squarely against them? The answer to that is: no, actually. The trades blow often but not constantly. In New Caledonia, for example, they blow only 80% of the time during the austral summer and 40% of the time in winter. At other times of the year, there are calmer periods with winds from the north, west or south. During such periods, one can sail to the east, even in a canoe that can only sail 75° into the wind.

This is the best strategy in any case. It would be much more dangerous to sail out with the prevailing winds on an exploring trip, since you could not be sure of getting back if, after a week or two at sea, you had found nothing. The risk is less if you leave with the wind from the beam, and non-existent if you sail out in the direction opposite that of the prevailing winds, i.e. towards the east. This is probably the tactic the Polynesians adopted. As it happens, since the trades blow from the ESE, they first colonized in an easterly direction (the safest), then towards the north (Hawaii), and finally the south (New Zealand).

When they discovered a new island, the colonists kept in contact with their island of origin, at least for a while. Both archaeology and the Polynesian oral tradition substantiate this. The Hawaiian sagas mention the period of the "great voyages" that lasted for several centuries, during which time Hawaii was in contact with Tahiti (a 4-week voyage in each direction). And how did they navigate? We will examine that subject in the next question.

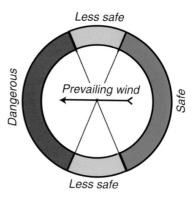

161. How did the ancient Polynesians manage to sail over such vast distances without compass or sextant and not get lost?

If there is a "race" of great navigators anywhere on Earth, it is the Polynesians. With no coastlines to follow and immense distances to cover, they accomplished their extraordinary feats thanks to navigational techniques that combined astronomical sighting and the recognition of precursor signs of land.

The sky, i.e. the apparent positions of the stars, changes as one moves north or south on Earth. To return to their home islands, the ancient Polynesians had to find "the island's sky" again (in modern terms, we would say "return to the same latitude"). The method they used for this was to find "the star hanging over the island," i.e. the star that passed over it at its zenith. In order to return to Hawaii from Tahiti, for example, they had to sail north until the star Acturus

(*Hokule'a* in Hawaiian[5]) passed at the zenith (directly overhead). Tahiti's zenith star was the constellation we call the Southern Cross. Fiji's star was Sirius.

Now if a zenith star indicates latitude, it gives no information at all about position in the east-west direction. This is the same problem that European navigators had before the invention of the marine chronometer: no information on longitude (Q. 159). The method the Polynesians seemed to use systematically was to take a route that would bring them to the proper latitude *upwind* of the target island. Then, with their zenith

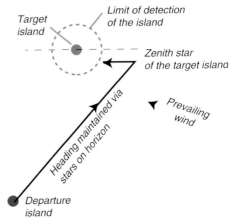

The principle of Polynesian naviga-tion.

star in place, overhead, all they had to do was come about and let the wind carry them to the island.

The Polynesian navigators had a very good mental map of their region, and they found the heading they wished to maintain thanks to stars near the horizon. They were lucky there in that, since they mainly sailed within a band between 20° from the equator, the sunrise and sunset azimuth (position on the horizon) of a given star is not much affected by latitude, usually not more than a few degrees [61].

Thus they had at their disposal a veritable "star compass," as precise as any magnetic compass, on condition that they memorize the positions of several hundred stars. For, since the Earth turns, one has to know the entire series of stars that rise or set in any one direction. And this star-based information was only available at night. In the morning and the evening they used the position of the Sun, as long as it was low

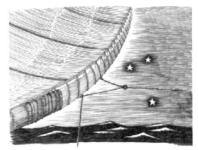

Guide stars at the horizon al-lowed the heading to be main-tained.

enough on the horizon. During the rest of the day, they took their cues from the direction of the swell and the wind.

---

[5]Whence the name of the famous replica of the Hawaiian seafaring canoe, which made several long passages in the Pacific, starting in the 1980's, using the ancestral methods of navigation.

In calm seas, from the stable platform that a double canoe can provide, the verticality of the zenith star can be measured to within about half a degree, or 30 nautical miles. The seafarers had to be able to detect land at that distance, then. For this, they read subtle signs in the clouds, the birds, the waves.

Clouds can often be seen stationed above Pacific islands. This can be due to the evaporation of a lagoon's relatively warm waters or to the presence of a mountain that forces the wind to rise and causes the water vapor it contains to condense. Even if the sky is full of clouds, those of local origin can be recognized because they are *stationary*, whereas ordinary clouds are moved about by the wind. Another possible clue is that clouds above an island can be tinged underneath with a slight color that reflects the color of the island or atoll beneath them (Q. 70).

The navigator has a friend in birds, too; they can indicate which way land lies over 20 miles from a coast. We are not speaking of confirmed seabirds here (their presence is quite unhelpful), but of birds like frigate birds and sea swallows that sleep on land. And even there, the direction of their flight can be misleading for neophytes; their behavior during the day tells you nothing. It is early in the morning when they fly out to go fishing, and again in the evening when they return to their nests, that the direction of their flight shows where land is to be found.

The third type of clue can be read in the crests of the waves and the direction of the swell. When the swell encounters an island, the wave crests change direction and form a chop downstream due to interference. Part of it is reflected off the coast upstream, and moves in the opposite direction. These are subtle clues, but the Polynesian navigators were (and some still are) past masters in interpreting them.

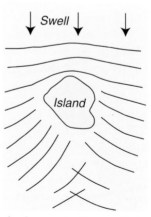

A *change in the swell indicates a nearby island.*

So if these masterful seafarers dared sail the vast stretches of the open Pacific, it is because they were armed with a prodigious store of sailing know-how. They had learned to read the sea and the sky as no other people before them. The ocean was not a barrier for them, it was a broad "highway."

162. To determine our position on Earth, why is it not enough to simply observe the sky?

When we travel north or south on Earth, from points A to C on the figure below, for example, the appearance of the night sky changes. A given star rises or sets in a different place on the horizon (azimuth) and culminates higher or lower in the sky. The stars that were passing directly overhead (at zenith) are replaced by new ones. Similarly, in daytime, the Sun culminates higher or lower, and rises and sets at different azimuths.

On the other hand, if one travels in the east-west direction, from A to B, for example, the sky itself will be unchanged, only the time of day or night when it looks the same will be different because of the Earth's rotation.

Observing the sky can thus inform us of our position in the north-south direction, i.e., in latitude, but cannot *by itself* give us any information about our position in the east-west direction, in longitude. If we want to know our position with respect to a reference point at the same latitude, we need to know how much (through what angle) the Earth has rotated between the time we made our observation and the time when the sky looked the same at the reference point. In other words, we need to know the *local time* at the reference point when the sky looked the same. For example, when the Sun is at its highest point in the sky, it is noon for us. If we know that, at that precise instant, it is 4:00 pm at Greenwich, our longitude is 4 hours to the west, or 60° W.

The simplest solution would be to take along a clock set to show the time at a reference meridian, such as the Greenwich meridian. But this clock would have to keep very accurate time. To know our longitude to within 30 nautical miles, for example, it would have to be accurate to within 2 minutes, i.e., for a six-week voyage, it would have to gain or lose under 3 seconds per day. Before the development of the marine chronometer by John Harrison in England, such accuracy was unattainable. Clocks of that era gained or lost up to a minute per day because their pendulums were overly sensitive to ship motion and temperature changes.

Harrison solved that problem by replacing the pendulum with a spiral spring. And, to compensate for the temperature effects, he used metals of different expansion coefficients. These two critical innovations are still used in mechanical watches today.

Harrison's H4 chronometer, completed in 1759 after 25 years of effort, could keep time to within 1 second per month. But these clocks were expensive and production was slow. Not until early in the 19th century did the marine chronometer come into common use at sea.

An alternative to using a mechanical clock is to use an "astronomical clock," i.e., rely on observing celestial events that take place so far away that, no matter where we are on Earth, they are seen at exactly the same moment. Eclipses are such events. If an

Harrison's H4; 13 cm in diameter it weighs about 3 pounds.

eclipse has been predicted to occur at the Greenwich observatory on a given day at 11:00 am, and if, on our way to the Caribbean, we see it occur at 8:00 am, our difference in longitude is 3 hours in time, i.e., 45° westward. Solar and lunar eclipses are not frequent enough for this method to be practical, but there are eclipses that occur much more often: the eclipses of the satellites of Jupiter by Jupiter itself. These satellites rotate with such perfect regularity that they constitute a marvelous clock, usable every clear night everywhere on Earth. One of the brightest satellites goes behind Jupiter roughly every 12 hours. Galileo, who discovered these satellites in 1609 with his new telescope, had proposed this solution and established the necessary tables. The method was indeed used with success for calculating longitude on land, but it was never used at sea due to the difficulty of pointing a telescope from the moving deck of a ship.

Another astronomical clock is supplied by our Moon. The Moon rotates around Earth in 27.3 days and therefore moves quite rapidly across the sky with respect to the stars: it covers a distance equal to its own diameter in one hour. If a table showing the position of the Moon with respect to a set of stars is established for a given point on Earth, Greenwich for example, the time at Greenwich can then be calculated at a different point on Earth simply by measuring the distance of the Moon to those same stars. This is the method of *lunar distances* used by Cook, Bougainville and other great navigators of the 18th century. But the calculations for this method are complex and it was not adopted by the majority of navigators of the time.

In the end, the winner was the marine chronometer, an instrument that remains the pillar of celestial navigation today.

163. How does a sextant work?

The sextant is the instrument that the GPS has replaced, much to the relief of the uninitiated who regard it with a mixture of respect and incomprehension. Although indeed somewhat complex and intimidating to use, its function is quite simple: it measures an angle.

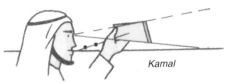

The ancestor of the sextant was the kamal, a rudimentary instrument used by Arab navigators from the time of Sindbad the Sailor. Used to find a predetermined latitude or to maintain a course on one for latitude sailing (Q. 159), it consisted of a small wooden plate with a knotted string attached. The string was held between the teeth and stretched taut, while the bottom of the plate was aligned with the horizon and its top with the North Star. The length of the string determined the proper angle for the desired latitude, and this length was itself determined by choosing the proper preset knot to hold between the teeth.

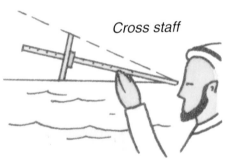

Cross staff

The cross staff (also called Jacob's staff) is a variation of the kamal which allows angles to be measured with better accuracy. It was this instrument that the European navigators used during the Age of Discovery.

The problem with these ancient systems for measuring angles is that one has to sight in two directions at once: towards the horizon and towards the celestial body. This is especially difficult from the deck of a moving boat because, as one's gaze moves from one direction to the other, the initial alignment is easily lost. In theory, the astrolabe solves this problem by acting as a pendulum to supply the vertical (instead of using the horizon): only one direction has to be sighted. On land, this became the instrument of choice, but its use is very imprecise at sea because of the boat's movement.

It was John Hadley who found the solution in 1730, thanks to an ingenious optical system *combining* the two directions to be sighted. The horizon is sighted through a semi-transparent mirror onto which the celestial body is projected by means of a second mirror.

The second mirror is movable and adjusted to align the celestial body with the horizon. When this is done, the height (altitude) of

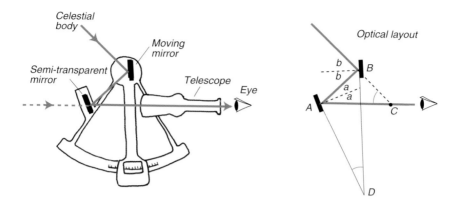

the celestial body (angle C in the figure) is given by twice the angle between the two mirrors (angle D).[6]

Hadley's instrument was actually an *octant*, with a range of 45°, that is to say 90° on the sky. This permitted the measurement of all distances between the horizon and the zenith, but the range was extended to 120° on the sky, i.e. 60° on the instrument (hence the name *sextant*, a 6th of a circle), in order to allow measurement between the Moon and stars by the lunar distance method (Q. 162). Although the lunar distance method was eventually abandoned, the sextant was retained rather than reverting to the octant.

164. Why do so many of the stars used in navigation have Arab names?

About 50 of the most brilliant stars in the sky are used in navigation, and all of them had been given names in ancient times. But after the fall of the Roman Empire and for centuries thereafter, scientific astronomy in the western world continued to be practiced only by the Arabs. Through their contact with the Greco-Roman world, Romanized Egypt in particular, they had become the heirs to and guardians of the science of the ancients.

The science that Europe rediscovered during the Renaissance was thus conveyed mainly through Arabic sources. In astronomy, for example, the works of Ptolemy, the last great ancient Greek astronomer, who lived in Alexandria around 140 A.D., were recovered

---

[6]In the triangle ABC one has $\widehat{C} = 180° - \widehat{B} - \widehat{A} = 180° - (180° - 2b) - 2a = 2(b - a)$, and in triangle ABD, $\widehat{D} = 180° - \widehat{B} - \widehat{A} = 180° - (90° - b) - (90° - a) = b - a$, hence $\widehat{C} = 2\widehat{D}$. The scale for $\widehat{D}$ is graduated to give $\widehat{C}$ directly.

only in the Arabic translation, the *Almagest*.[7] This is why most of the brilliant stars have kept their Arabic names.

Many of these names have poetic meanings as you can see from the examples in the table below, where the common names and Arabic meanings are given for a few of the stars often used in navigation [1].

Name	Arabic name	Meaning	Scientific name
Achernar	Akhir an-Nahr	End of the river	α Eridanus
Aldebaran	Ad-Dabaran	The follower (of the Pleiades)	α Taurus
Altair	At-Ta'ir	The flying eagle	α Aquila
Betelgeuse	Yad al-Jauza'	Orion's hand	α Orion
Deneb	Dhanab ad-Dajajah	The hen's tail	α Cygnus
Eltanin	At-Tinnin	The big snake	γ Draco
Rigel	Ar-Rijl	The foot	β Orion
Shedir	As-Sadr	The breast	α Cassiopeia

The positions of the stars used for navigational purposes are listed in the nautical ephemerides, or nautical "almanac." This word, too, comes from the Arabic, *al-manakh*. Originally the title of a compendium of geographical and climate data, the word eventually came to mean a list of celestial data. The ephemerides were first officially published by the Paris Observatory in 1696. The first British *Nautical Almanac* came out in 1767.

It is obviously not feasible to give every star in the sky a common name, so professional astronomers refer to stars and other celestial objects by a system of codes. The system most commonly used today for the brighter stars was proposed by Johann Bayer in 1603. It consists of a code name for the constellation in which the star appears preceded by a Greek letter (or a number, if one has run out of letters) indicating the star's rank, in brightness, within that constellation. For example, the scientific name for Adelbaran is α Tau (the brightest star in the constellation Taurus, the Bull), and the brightest star in the constellation Centaurus is the famous α Cent (Alpha Centauri), the star that is our closest neighbor.

---

[7]The Arabs were so impressed by Greek science and by Ptolemy's work in particular that they called his book *Al Magiste*, meaning "the greatest," a term which eventually evolved into Almagest.

165. Why is there a difference between magnetic north and geographic north?

The magnetic needle of a compass almost never points towards the geographic, or true, north. Christopher Columbus noticed this effect during his Atlantic crossing. And the Portuguese seafarers circumnavigating Africa in the 1490's had probably noticed it, too: at the southern tip of the continent, unlike other places on Earth, their compass needles did (at the time) point towards the geographic north, and some say that this is why they called it the *Cabo das Agulhas*, the Cape of Needles.

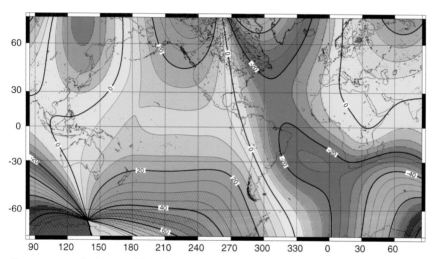

Compass variation on Earth in 1995. Variation can be as great as 20° west (green) or east (brown). Source: *USGS.*

The reason for this discrepancy is that Earth's magnetic poles do not coincide with its geographical poles (the points through which our planet's axis of rotation passes). The two points are actually quite far apart.

The north magnetic pole is currently located in the Canadian Arctic at a latitude of 81° north, about 500 miles from the geographic north pole, and it is moving ever further north by about 25 miles per year. The south magnetic pole is located in the Antarctic Ocean at a latitude of 64° south, about 1,500 miles from the south geographical pole.

The difference between the positions of the magnetic and geographical poles is called *declination* or *variation*. Anyone still sailing by compass these days (or if the GPS goes out) must also take care not to confuse declination with the *deviation* of the compass, which is the difference between the north indicated by the instrument on the boat and true magnetic north. This difference, which is caused by local magnetic effects in the vessel itself, is corrected by installing magnets and iron spheres close to the compass (Q. 166).

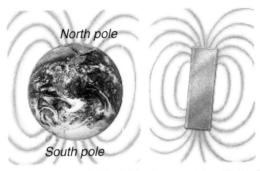

North pole

South pole

The magnetic field of Earth resembles that of a bar magnet.

The magnetic field of Earth is similar to the one that a bar magnet would produce if placed inside our planet: lines of force arc between the north magnetic pole and the south magnetic pole. There is a lot of iron in the Earth's interior, but that cannot be the source of our magnetic field because the temperatures in the interior of our planet are so high that all magnetized substances lose their magnetism.[8] So why the Earth has a magnetic field at all is something of a mystery. Scientists now think that it could be produced by convection movements in the viscous masses inside the planet acting as a dynamo and creating the weak field we detect at the surface.

These movements of the semi-liquid substances inside Earth are postulated to be imperfectly concentric with the planet's axis of rotation, thus giving rise to the discrepancy between the magnetic and geographic poles. The location of the liquid's axis of rotation varies over time, and the magnetic polarity can even reverse itself. This happens on average about once every 300,000 years, but the last time it happened was 740,000 years ago. That makes some geologists think that we are more than a little overdue...

When magnetic polarity does reverse itself, it happens rather quickly, over a period of just a few thousand years. While this is going on, the Earth temporarily loses the protective magnetic shield that normally deflects the Sun's lethal radiation, with potentially tragic consequences for the animals and plants of our world.

The value for declination is given on nautical charts, but you should not make the corresponding heading correction if you are using a GPS, which automatically corrects for the discrepancy, taking into account the boat's position on Earth and the date.

---

[8]The temperature at which a substance loses its magnetism is called the *Curie point*; the Curie point of most substances is less than 1000 °F, which is reached only about 20 miles beneath the Earth's surface.

166. What is the purpose of the two iron spheres found on the binnacles of large ships?

Iron compensation spheres.

Commercial vessels are constructed of iron plates. Although not very magnetic, they sit stationary for months during the construction period and end up being slightly magnetized by the Earth's own magnetic field (by sitting in a magnetic field, metals acquire their own magnetism, an effect known as *induction*) and by the hammering and vibrating they are subjected to. After launch, the ship will be battered by the waves and slowly become demagnetized, though never completely so. There always remains a low level of what is called *permanent magnetization*, and this residual magnetization has a specific, fixed direction and must be compensated for. This is easily accomplished by installing small magnets near the compass.

Besides this slight permanent magnetization with fixed orientation with respect to the ship's axis, an iron ship acquires an additional magnetization induced by the Earth's magnetic field. But this new "layer" of magnetism strengthens, weakens, disappears and changes orientation in direct response to the strength and direction of the magnetic field it encounters, which itself varies with the ship's course and its position on Earth. Corrector magnets cannot solve this problem because their effect is constant, and the correction needs to change as the field changes. The solution is to install soft iron spheres near the compass to counteract the influence of the ship. Since the ship's mass resides essentially in its length, the iron balls are placed perpendicular to it. Their distance from the compass is then adjusted so that the magnetism they acquire from Earth's field exactly cancels out that of the ship's hull.

167. Why is Greenwich Mean Time (GMT) also called "Zulu" time?

In the past, time was defined locally, by the Sun, usually with the help of a sundial. Consequently, the time of day varied from town to town as one traveled in the east-west direction overland, but the slight, incremental differences between neighboring places seldom mattered because travel was quite a slow affair. But with the advent of railroads, time discrepancies became a nightmare.

By 1840, the British railroads had decided to adopt one single, official time for their entire network. That was obviously impossible for the United States, however, with a territory covering 55° of longitude, 3.5 hours of time difference. But how could travelers know how long a trip was going to last if the time shifted with every mile traveled? How could you write a train schedule? How could you coordinate train connections for different legs of a journey involving different railroad companies unless both companies ran on the same time system? So in 1883, at the instigation of the American railroad companies, the United States decided to officially divide the country into four time zones. And the following year, an international conference meeting in Washington, D.C. extended the system to divide the entire world into 24 zones, each one 15° of longitude wide. The meridian of Greenwich, England was adopted as the reference point, with

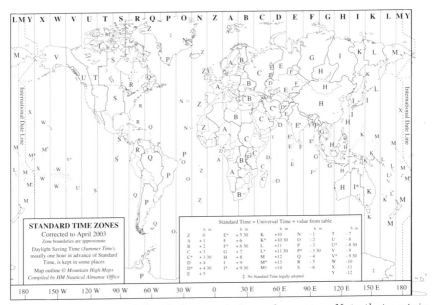

The 24 time zones and the choices made by each country. Note that certain countries have adopted a 1/2 hour step. Source: *H.M. Nautical Almanac Office.*

the mean solar time of Greenwich (Greenwich Mean Time, GMT) furnishing "time zero."[9]

In civilian use, the time zones are designated by a code usually composed of three letters. The zone containing Greenwich is GMT (Greenwich Meridian Time) or WET (Western European Time). The 48 lower states of the U.S. have EST, CST, MST and PST (Eastern, Central, Mountain, and Pacific Standard Time), east to west.

The military, especially navies, prefer to use the letters of the alphabet for different zones, easier for radio and even written communications. The zones run from A through Z, omitting the letter J (25 letters are needed rather than 24 because the zone covering the international date line is divided into two). The zone containing the Greenwich meridian was given the letter Z for *zero*, which, in the international phonetic alphabet is "Zulu." So the time zone designations GMT and Zulu are equivalent.

168. What is the difference between UT, UTC and GMT ?

For celestial navigation and for astronomers, the only time that really counts is the one tied to the rotation of the Earth. This used to be Greenwich Mean (or Meridian) Time and is now replaced by *universal time*, or UT.

Universal time is determined by observing the meridian passage of the Sun or of a star of known position. But because the Earth's rotation is slightly irregular and is also slowing down, UT varies by about a millisecond per day. And since the atomic clocks that we now have can provide even more precise timekeeping than astronomical observations can, we may as well use them. Thus, Coordinated Universal Time (UTC)[10] was born, based on atomic timekeeping that fluctuates not one jot from day to day. It has become the basis for civil time in all countries. In order to keep civil time in synch with the universal time that is linked to Earth's rotation, UTC is adjusted to coincide with UT when the accumulated difference reaches a full

---

[9]This decision was a mighty disappointment to the French, who had hoped to have the meridian of Paris adopted. But the choice was almost inevitable since, at that time, 75% of all existing nautical charts were already based on the Greenwich meridian. So the French gave in but asked for a promise in exchange: that the English-speaking world would adopt the metric system. A hundred years have gone by and the French are still waiting.

[10]Normally, Coordinated Universal Time should give the acronym CUT, but that was felt to be inappropriate.

second: as it turns out this means that, every year or two, a second is inserted at midnight at the end of June or December (in the same way that a day is added to the calendar in leap years). Someone doing celestial navigation can use UT and UTC interchangeably if precision to within one second is acceptable to them. For greater precision, correction coefficients have to be applied to UTC to get UT. These coefficients are broadcast regularly.

The National Institute of Standards and Technology (NIST), an agency of the U.S. Department of Commerce, broadcasts UTC continuously with its short wave radio station WWW located in Fort Collins, Colorado (on 2.5, 5, 10, 15 MHz and 20 MHz). UT is not broadcast. Your GPS can also furnish UTC, but only after it has been running for a while. The GPS transmitters are linked to an atomic clock that is synchronized with the clocks used to define UTC but do not register the seconds that are inserted annually or semiannually (in 2006 the discrepancy had reached 14 seconds). When you turn on your GPS, it initially shows GPS time. The satellites broadcast a message containing the value of the discrepancy every 12.5 minutes. When a GPS receives this message, it corrects the GPS time and only then begins showing UTC time.

The use of GMT time is being discouraged, since the more precisely defined UT and UTC have replaced it. But old habits die hard, and you will still hear people using GMT as a synonym for UT or even UTC.

169. When entering a port, you leave the red buoys to starboard in America and to port in Europe. How in the world did that come to be?

It's true! Throughout the Americas, including the islands of the Caribbean, even if they happen to be English, French, or Dutch possessions, the rule is *"Red, Right, Return."* But when you sail into a European port, be it English, French, Dutch or any other, you must leave the red buoys on your port side. Crazy as it seems, the world is divided into two zones, *A* et *B*, with the beaconing in one being the reverse of the beaconing in the other. Generally speaking, zone *A* is historically the zone of British influence, while zone *B* reflects American influence.

The situation is incongruous: marine buoys are international by nature and should be identical everywhere on the planet. And yet, if the situation is what it is today, it is certainly not for lack of debates and international conferences.

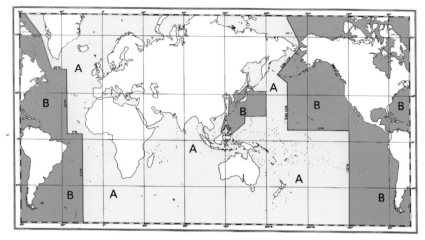

The International Association of Lighthouse Authorities (IALA) system of buoyage: areas A (blue) and B (orange).

The first international conference on the subject was held in 1889 in Washington, D.C., and discussions went on for two months. The decision was made to adopt a lateral system based on colors and shapes: a black cylinder portside for ships coming into port, and a red cone to starboard. But at the time it was just a question of paint colors, since buoys and towers were not yet equipped with lights.

The same conference in Washington also examined the rules designed to prevent collisions at sea and decided that vessels should be equipped with running lights for navigation at night. It would have been sensible to rule that the ships' lights and the buoys should be the same color, making the night and day signals identical; starboard lights would be red, for example, like the red buoys left to starboard as one sailed into port. But the issue was settled differently: the color red was adopted for a ship's portside light, green for the starboard light. That led to today's "schizophrenic" situation.

An attempt was made to correct the anomaly early in the 20th century, when buoys began to be equipped with lights. But in spite of a whole series of international conferences, there has been no meeting of minds. The Europeans, especially the British, have wanted to adopt the "ships' lights" rule: when two ships meet in a channel, each stays right, giving red light against red light. Hence it seemed best to have red channel lights on the port side of a harbor entrance so that a ship's pilot would react the same way to a red buoy or jetty light as to the red running lights of another ship. Americans, on the other hand, have wanted to uphold the decisions of the Washington conference. The two opposing viewpoints were cemented into the international rules at a conference in Lisbon in 1931. The world has been divided in two ever since.

170. Why were the colors red and green chosen for running lights?

The decision to use red and green running lights on ships was made in 1889 (Q. 169). Since then, with traffic lights on land having adopted the same colors, red and green have come to seem so natural that we can hardly imagine using any other colors. But, looking back, couldn't some other color combination have served equally well? Wouldn't blue and yellow have been more highly visible, for instance?

In scientific terms, there is nothing fundamental about color at all. There is no blue, red or green, not even black or white, in the absolute sense. Photons and light waves exist, but the impression we have of color is simply the way our brains have evolved to interpret light with a wavelength between 400 and 700 nanometers (a nanometer, abbreviation nm, is a billionth of a meter). White is nothing but the perception of all the colors at once, i.e., the light of our star, the Sun, between 400 and 700 nanometers that penetrates our atmosphere. Black is the absence of light waves in this bandwidth.

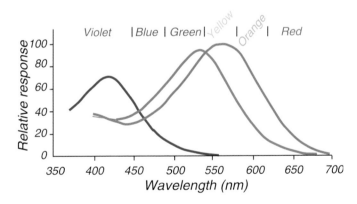

Response curve for blue-, green-, and red-sensitive cones in the human eye.

Our eye evolved to discern colors so that we could understand the world around us better. Color helps us detect what we need — and need to avoid — for survival. Color vision is certainly more informative than seeing only in black and white would be. Our eye is capable of discriminating 10 million shades of color, but, astonishingly, we have only three different types of "detectors" to work with:

our blue-, red-, and green-sensitive cones. Each of the three types is sensitive to quite a wide band of wavelengths, but is most sensitive for certain ones: 420 nm for blue-sensitive cones, 530 nm for the green, and 560 nm for the red. The red ones are actually misnamed: they are really most sensitive to the color yellow. Our sensitivity to red is five times less than to yellow or green. This is easy to confirm at sea: for lights of equal intensity, we can detect green ones long before red ones (at approximately twice the distance).

So why choose green and red for our color signals? Green is justified because the eye is so sensitive to it. As for red, there it is a question of contrast. Green and red are different enough in wavelength to prevent any possible confusion. Red is actually the complementary color of green.[11] Note that this only holds true for lights. By day, green buoys are harder to see against the blue-greens of sea and sky, whereas red stands out much better.

171. When were the first lighthouses built?

The practice of using lights to guide sailors to home port at night is certainly very ancient. It is mentioned in the Odyssey, as a matter of fact. These first "lighthouses" were just simple bonfires set on high ground or atop watchtowers.

The lighthouse of Alexandria.

The most famous lighthouse of antiquity was the one built in 300 B.C. on the island of Pharos, at the entrance to the port of Alexandria. One of the seven wonders of the ancient world, it was about 300 feet high and its light could be seen 25 miles away. It was the second tallest monument in antiquity (the great pyramid of Giza was the tallest). In 400 A.D. there were over 20 lighthouses in the Roman Empire, between the Black Sea and the English Channel.

The lighthouse of Alexandria stood for over 16 centuries but finally collapsed in 1303 during an earthquake. The lighthouse of Boulogne on the English Channel, built in the first century A.D., remained in service until 1644 when the cliff it stood on crumbled

---

[11]That is, if red and green are combined, for example by spinning a disk that is half red, half green, the eye sees white. This is a purely physiological effect: the two colors, seen simultaneously, stimulate the eye and brain the same way that white light does.

The principal lighthouses of the Roman Empire.

under it. Very few new lighthouses were constructed during the Middle Ages, and it was not until the end of the 17th century that a serious effort was made to light the coastlines of the major European countries. The oldest lighthouse in North America is the Tulum lighthouse in Mexico, which in pre-Columbian times guided Mayan mariners through the Caribbean Sea. It dates from the 13th century.

172. If a lighthouse bulb only puts out a few hundred watts, what makes its beam so visible so far?

The lighthouse is an object of near mythical stature, a noble sculpture embodying victory over the blind violence of nature, and man at the service of his fellow man. We remember the dramatic images of lighthouses exposed to furious seas, the great difficulties overcome during their construction on rocky, fogbound islets, the devotion of the men who tended them, keeping the lamp lit... But the true beauty of a lighthouse is really in its *optics*. Isn't it extraordinary that a lighthouse equipped with a lamp only a few hundred watts strong can be seen up to 20 miles away? Of course it is true that, in the dark of night, the tiniest light can be seen quite far off. Henri de Monfreid, the dashing French adventurer, relates how the Arab rebels waiting on shore for a night delivery of weapons from him used the discreet glow of a lit cigarette to signal which cove to pull into. Still, how can an almost standard light bulb be seen so many miles away?

Plain wood fires were used in early lighthouses in Europe, then, later, coal fires. Starting in the 1700's, the fires were protected from wind and weather by placing them under domes or surrounding them with glass panes. But the fires consumed huge amounts of wood or coal that had to be brought up to the top of the tower, and then the keepers had to feed and tend the flames all night.

Open fires were eventually replaced by oil lamps installed in the focal plane of a parabolic reflector, and a glass "chimney" was added to improve the draw and keep smoke from sooting up the reflector. The wick absorbed the oil automatically, by capillarity or with the help of a small pump, and the nightly duty of the keeper was essentially reduced to moving the wick up from time to time.

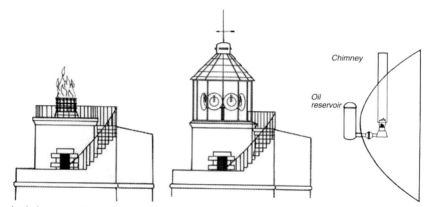

At left, a medieval lighthouse with a wood or coal fire; center, a lighthouse with multiple oil lamps and reflectors; at right, a detail of one of the lamps.

But now there was a big new problem to solve: an oil lamp really does not emit much light. Sailors of a traditional bent know about that! A brass oil lamp looks oh-so-nautical swaying on its chain, and it transforms the cabin with its warm golden glow, but just try to read a book by it! It supplies less light than a 5-watt electric bulb.

The problem is that the luminosity[12] of a flame is about 1000 times less than that of an electric filament. This is because the luminosity of a body depends on its temperature: the higher the temperature, the greater the luminosity. An electric filament reaches a temperature of about 3500° F, while an oil flame burns at about 200° F. An electric lamp can therefore get along with a tiny filament, while a gigantic wick would be needed to produce the same amount of light by burning oil.

To make up for the weak luminosity of the flame, the size of the wick was increased as much as possible, and several wicks were even installed side by side. But the size of a light source cannot be increased indefinitely. When the surface of the source becomes

---

[12]The luminosity of a light source is the amount of light emitted per unit of surface area.

much larger that a simple point, the light emerging from the reflector or lens is no longer in the form of a parallel beam. It spreads out, its light energy diluted in a widening cone rather than staying in a tightly concentrated beam. As a result, the distance over which the beacon is visible does not increase.

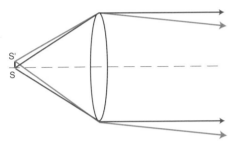

If a point, S', inside a light source is off axis for a lens, the beam of light that it emits will travel in a direction different from that traveled by a beam emitted by an on-axis point, S.

The power of a beacon light depends on three factors: the luminosity of the light source, the efficiency of the optics, and the size of the optics. Since everything that could be done had already been done to optimize the first factor, further progress would have to come from improving the two remaining factors.

The efficiency of a lighthouse system is equal to the amount of light coming out of the optics (reflector or lens) compared to the amount of light produced by the source. Reflectors were not very good in the late 18th century, and their surfaces tarnished quickly. Efficiency was scarcely 50%. Hence, glass lenses soon replaced reflectors.

It was then time to work on the third factor, the size of the optics. The importance of a large lens is easy to understand: the larger and closer to the source it is, the more light it can capture from the source. The problem is that a large lens has to be thick. That makes it heavy and it also absorbs some of the light passing through it. Worse yet, a large lens placed very near a source[13] works poorly: it suffers from "aberrations," meaning that the beam emerging from it is not tightly concentrated, but diverges into a wide cone.

It was Augustin-Jean Fresnel, the great, early 19th century French physicist, who found the solution to this problem. To minimize the thickness of the lens and eliminate aberrations, he invented (in 1822) a system composed of a central lens with surrounding concentric prisms, which has been called the *Fresnel lens* ever since. This lens improved on the system based on mirrors by a factor of about 300 times!

Augustin Fresnel.

---

[13]In optics, this is called a "very open" lens. For a camera, it would be the equivalent of an objective open at f/0.5.

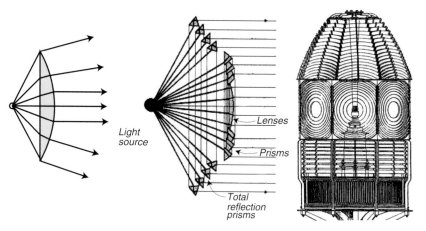

A very "open" single lens cannot produce a parallel beam (left). Fresnel's solution consists of surrounding a lens with circular prisms to capture more light from the source (center). At right, an 8-panel Fresnel lens.

The results were revolutionary. Soon afterwards, most lighthouses worldwide were equipped with this type of lens. The light system had become so efficient that, when electric lamps eventually replaced the mediocre oil lamps, a simple 100 to 250 watt bulb usually sufficed.

Today, thanks to progress in electric lighting, high-luminosity lights are readily available (arc lights or halogens), but Fresnel lenses continue to be used because of their excellent efficiency. The lenses are produced in molds now, rather than by the laborious and expensive optical polishing method once used.

173. How are the lights on offshore buoys powered?

Seeing a light on a buoy tethered far from shore, sometimes in really deep water, does make one wonder — where does the power come from?

The earliest buoy lights were powered by acetylene gas. The buoy walls had to be thick enough (more than 1/2 inch) to handle the high pressure. Propane was used in the 1980's, and this gas, too, was stored inside the body of the buoy. All buoy lights are now powered electrically, via solar-powered batteries. The bulbs, either halogen lights or sets of electroluminescent diodes (LED's), provide only 5 to 10 watts, but this is enough to be visible from 3 to 5 miles away.

174. Why is visibility better after a rain or when it is cloudy?

Suppose we start with the term *visibility*. For any object to be visible, the important thing is not the luminosity of the object itself, but the *contrast* between the object and the background. A lighted candle in a sunny room produces little effect, but if the room is darkened, that same candle seems brilliant. Likewise, we can see the stars clearly on a dark night, but if the Moon rises, only the brightest stars remain visible. The sky background has become too bright. For an object to be visible against its background, its luminosity must be at least 2% greater than the luminosity of the background.

During the day, the background light is due to the fact that the air itself is luminous. Molecules of air and the dust and water droplets suspended in it scatter sunlight in all directions, and some of that scattered light lands on our retina (Q. 66). Not only does the sky at the horizon thus become brighter, but this glow of "air light" creates a sort of veil between us and the object.

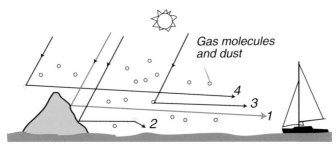

The visibility of a distant object that reflects light (ray 1) is reduced because part of that light is absorbed by the atmosphere (ray 2), because there is scattered sunlight (air light) between us and the object (ray 3), and because of the luminosity of the sky behind the object (ray 4).

Visibility improves after a rain because the raindrops, as they fall, pull down with them the suspended solid particles and small water droplets, leaving the air very clear. The scattering of light by the atmosphere is then greatly reduced, and visibility is excellent.

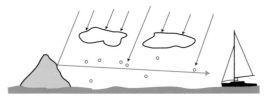

When there is a partial cloud cover, the clouds block some of the Sun's rays that would normally strike air molecules between us and the object. This reduces the glow of air light, improving visibility [44, 49, 6].

175. Why is it easier to see the light from a lighthouse at night than the lighthouse itself by day?

A lighthouse can only be seen by day if the contrast between it and its background (sky or landscape) is great enough. The amount of *air light* (luminosity of the air itself) is the main factor that reduces this contrast by day (Q. 174).

At night, air light disappears and the visibility of the lighthouse beacon is only limited by the purity of the air and the intensity of the light beam. Generally, beacons are designed to be visible at a greater distance at night than the lighthouses would be, under conditions of good visibility, by day.[14] When atmospheric conditions are good, daytime visibility is usually about 10 nautical miles. A lighthouse with a luminous range of 20 nautical miles would be visible at that distance by night, but the lighthouse building would not be visible at that distance by day.

176. Why does a lighthouse beam end abruptly in mid air, rather than fading slowly away?

It really is surprising the way a light beacon seems to stop short, as though cut off or blocked at some point along its length, rather than just growing ever dimmer with distance. A strong flashlight or searchlight can produce the same effect.

We must remember here that a beam of light, when viewed from the side, is only visible to a viewer because the air through which it passes contains dust or droplets of water that scatter the light, and some of that light is scattered in the direction of our retinas (Q. 66). The more particles in the air, the more light is scattered in our direction and the greater the visibility of the beam. But the reason why the beam seems to end abruptly is certainly not because the air suddenly becomes free of particulates; it is a simple question of geometry, as the sketch below illustrates.

If you imagine yourself standing at point O, you could see the scattered light of the beam coming from directions A, B, C, etc. And the luminosity of the beam would not fade at all in the distance because its apparent thickness increases (the thickness of the beam

---

[14]By definition, visibility in daylight, called meteorological visibility, is the distance at which the contrast between an object and its background is reduced to 5% of the value it has when one is standing directly in front of it. Meteorological visibility is less than 5 nautical miles in light fog, and can be as much as 25 nautical miles when conditions are exceptionally good.

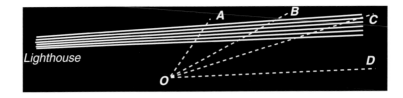

along the path OC is greater than along OA, which is closer). But even if the beam traveled very far, it would never be visible to you in the direction OD, parallel to the beam. For the observer at O, this direction marks the abrupt "end of the beam" [49].

177. Why is the Mercator projection used for nautical charts?

It is impossible to make a perfect representation of a spherical surface (which has 3 dimensions) on the 2 dimensional surface of a chart: distortion is inevitable, no matter what you do.

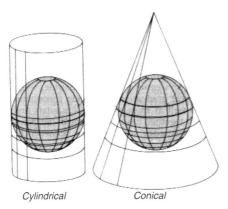

Cylindrical   Conical

The technique of representing a spherical surface on a plane surface is called making a *projection*. Projections are most commonly made onto a cylinder or a cone that is then unfurled, figuratively speaking, to make it flat. Projection onto a cone, which minimizes local distortion, is normally used for land maps covering small areas. Projection onto a cylinder from the center of the Earth seriously distorts the upper latitudes, but has the advantage of making the lines of latitude and longitude form a grid with right angles. But contrary to popular thought, that is not enough for navigation.

A navigator has to be able to easily determine the proper heading to follow when sailing to a given spot, and this is easiest to do if the route is a straight line when plotted on a chart. On Earth, such a route is a segment of a spiral, a spiral that cuts each meridian at the angle of the heading being followed. For the projection of this spiral to be a straight line on the chart, the spacing between

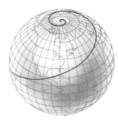

parallels must satisfy a special condition that the simple cylindrical projection from the center of the Earth does not.

The famous cartographer Gerhard Mercator found the solution in 1569 [14], arriving at the answer empirically, by trial and error. As the following figure shows, it consists of the projection on a cylinder from a point that changes as the latitudes changes. The mathematical formula describing the spacing of parallels was relatively complex for those times, and was only determined 30 years later.[15]

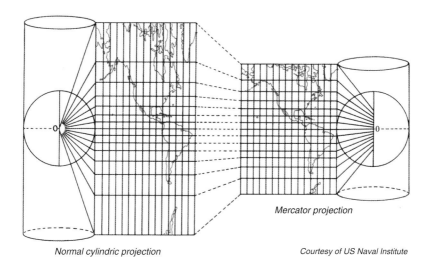

Mercator projection

Normal cylindric projection   *Courtesy of US Naval Institute*

One good way to think of a Mercator projection is to imagine the Earth as a balloon sitting inside a glass cylinder. Now, inflate the Earth-balloon until it touches the cylinder walls along the equator. If you continue to pump in air, more and more of the surface of the balloon will press against the glass cylinder, with the high latitudes being stretched more than those near the equator. The result is a Mercator map.

The Mercator projection was a fundamental discovery for navigation, permitting, as it does, the conservation of angles.[16] Not only

---

[15]By the British mathematician Edward Wright in 1599. The conservation of the angles implies that the scales for latitude and longitude must be identical *locally*. In the simplified case in which Earth is treated as a perfect sphere, if the meridians are to be parallel on the chart, the latitude scale must vary as $1/\cos\varphi$, where φ is the latitude (Q. 179). Since the longitude scale is locally identical, the positions of the parallels of latitude are thus given by $\int \frac{d\varphi}{\cos\phi}$. After integration, we find that the positions of the parallels on the chart are given by the formula $R\operatorname{Log}\left(\operatorname{tg}(\pi/4 + \varphi/2)\right)$, where R is the radius of the Earth and φ is the latitude in radians.

[16]A projection in which the angles are conserved is called *conformal* in mathematics.

does that allow headings to be read directly from the chart, but the angles measured with a bearing compass can also be transferred directly. As a result, the local scale is the same in all directions; it does vary with latitude, however (Q. 179).

The shortest distance between two points on Earth is a segment of a "great circle," and these do not intersect all meridians at the same angle. Sailing along a straight line on a Mercator chart is thus not the shortest route, but the difference with a great circle route[17] is negligible for distances under 1000 miles.

Although almost all nautical charts are based on Mercator projections, this is not true of aeronautical charts, as pilots do not usually navigate by compass but go point to point, following radio beacons. Since radio waves follow paths along great circles on Earth, flight charts use a special conic projection (called a Lambert projection) which is also conformal, but where the meridians are not parallel to each other and the great circles are indicated as straight lines.

178. Why is there a different value for a mile on land and a *nautical mile*?

There are no signposts or milestones at sea. To find out how far away something is, you can only rely on charts and the coordinates of latitude and longitude. The nautical mile, which is defined as one minute of arc at the surface of the Earth, is therefore an extremely practical unit.

The idea of establishing a unit with the length of a minute of arc at the Earth's surface is credited to Gabriel Mouton, an abbot from Lyon, France, who proposed it in 1670. Because this measurement was tied, not to any one country but to the planet as a whole, he hoped that people everywhere would agree to adopt it.[18]

Actually, since the Earth is more of an ellipsoid than a sphere, the length of an arc subtending a minute varies slightly with latitude (from about 1843 meters at the equator to 1861 meters at the poles[19]). So an average value would be used, typically the length of

---

[17]A route that follows a great circle is called *orthodromic*; a route with no change in heading is *loxodromic*; *loxos* means oblique in Greek.

[18]This idea of universality later led to the establishment of the meter, with a kilometer being originally defined as 1/10,000 of a quarter of the Earth's circumference. Mouton can thus be considered the father of the metric system, particularly as he had already proposed using a decimal system rather than fractions.

[19]According to the definition of geodetic latitude, the length of the minute of arc varies with the *radius of curvature*, not with the distance to the Earth's center.

a minute of arc at a latitude of 45°. But for many years, different countries used slightly different values: the American mile was equal to 1853.25 meters, the English, or *Admiralty*, mile was 1853.18 meters, and the French mile was 1851.85 meters. It was only in 1929 (1954 for the U.S.) that the length of the nautical mile was agreed upon internationally: 1852 meters *exactly*.[20]

1 Roman pace

The word *mile* itself comes from the Latin *mille passus*, "a thousand paces." For the Romans, a pace was not one step, but *two*, and it was equal to 5 of their foot units. That made one Roman mile equal to 5,000 Roman feet, a nice round number to deal with. It seems odd, then, that the English (and subsequent Americans) would have retained the Roman unit of the mile but changed its value to the not so nice-and-round 5,280 feet. The story behind that transformation is that, in medieval England, the main unit for measuring fields had become the *furlong*, which was the average length of a *furrow*, the distance an ox team could pull a plow before stopping to rest a bit as the plow was turned around and a new furrow was started. This distance eventually became fixed at 660 feet. Then to harmonize the furlong with the old Roman mile that was still used for English road distances, Elizabeth I decided by statute that the new mile would be exactly 8 furlongs, or 5,280 feet, in length.

179. Why must the latitude scale be used to measure distances on a chart?

By definition, one minute of arc on the Earth's surface equals one nautical mile. Hence, the number of miles between any two points on a chart is equal to the number of minutes of arc that distance would represent on a great circle passing through those two points. Meridians are great circles that are conveniently graduated along the side of a chart. All one has to do is transfer a distance measured with a divider onto the latitude scale to find the distance in minutes, hence in nautical miles.

But except for the equator, parallels are not great circles (i.e., centered at the center of the Earth). They are "small circles," and

---

The Earth being flatter at the poles, the radius of curvature there is larger, hence the longer length of a minute of arc there.

[20]There is no official international symbol for this unit. The nautical mile is often abbreviated N.M., or nm when there is no risk of confusion with nanometer.

the distance represented by one minute of a parallel (a longitude measure) will vary with latitude. The more one approaches the poles, the shorter a longitude minute will be.

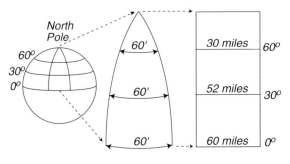

For this reason, distances must always be measured on the *latitude scale*. If the longitude scale is used instead, it could lead to sizeable error: 1° of longitude covers 52 nautical miles at a latitude of 30°, but only 30 miles at a latitude of 60°!

Particular care is required with charts having very small scales, showing the entire Atlantic, for example: the distances measured with a divider must be transferred to the latitude scale at the *same* latitude. This is because the Mercator projection used on charts does not preserve uniformity in the latitude scale (Q. 177). The difference is negligible on large-scale charts, but that is no longer the case for charts covering dozens of degrees of latitude.

180. Why is speed measured in *knots*?

Measuring speed with a line that has regularly spaced knots. Source: Musée de la Marine, Paris [58].

The custom of measuring a boat's speed in knots comes from the old method of throwing overboard a small plank attached to a line with regularly spaced knots tied into it. Speed could then be easily deduced by counting the number of knots that passed through one's hands in a given time.

One common method for measuring the time was an hourglass set for 15 seconds (1/240 of an hour). If the knots were positioned at 1/240 of a nautical mile, or 7.72 m (25.3 ft) apart, the boat's speed could then be read directly in nautical miles per hour.

181. How can the GPS be so precise?

The secret of the GPS's fantastic precision is that it uses *time* as the basis of measurement. With current technology, we can reach much greater precision in the measurement of time than in the measurement of distances or angles. As a matter of fact, the meter is now officially defined, not by a physical model meter, or standard, as used to be the case, but by the amount of time it takes for light to cover a certain distance.[21]

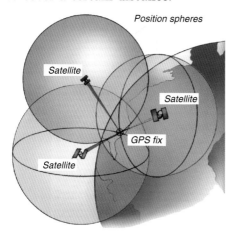

Position spheres

Satellite

Satellite

GPS fix

Satellite

The "GPS fix" is found at the intersection of the spheres of position corresponding to the satellites overhead at any given moment.

A GPS plots our position on Earth by measuring the time it takes for radio waves emitted by satellites to reach us. The distance to each satellite is simply equal to the time that takes multiplied by the speed of light (which is the same for radio waves and all other electromagnetic waves). Three satellites suffice to furnish our position. With measurements from one satellite, we learn that we are on a position point located somewhere on the "sphere" having the satellite at its center and a radius equal to the distance measured. With two satellites at work, we know that our position is somewhere on the circle formed by the intersection of the two corresponding spheres. And with three satellites, it is sure that our location corresponds to one of the two points common to the three spheres. One of those points is far from Earth and can be disregarded. Our position is the other point.

The GPS satellites are in orbit 11,000 nautical miles above Earth. There are 24 of them, which guarantees that at least three are overhead at any one time, no matter where one might be on Earth. If we want to measure the distance from each satellite to within, say, 30 feet (about 10 meters), and given that the speed of light is 300,000 km/s, we would need clocks with a precision of better than $3 \cdot 10^{-8}$ second, for the satellites *as well* as for the receiver.

---

[21]The current definition of the meter is the distance that light travels in a vacuum in $1/299{,}792{,}458$ ths of a second.

If, for example, a satellite sends a signal at time 1, and it reaches the receiver at time 1.7, the amount of time that the signal spent traveling can only be accurately known if the receiver clock is as accurate as that of the satellite clock and is synchronized with it.

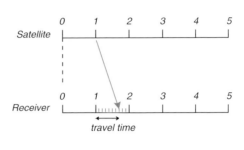

Each satellite is equipped with an atomic clock accurate to within 3 nanoseconds ($3 \cdot 10^{-9}$ s), and all these satellite clocks are synchronized among themselves. Unfortunately, atomic clocks are very expensive (each costs over \$100,000) and installing one in everybody's receiver was out of the question. But a solution was found, and an elegant one at that.

The clocks in the receivers are of the simple quartz type found in any cheap, modern watch, but they are converted into clocks with atomic precision by calibrating them with respect to the ones on the satellites. This is accomplished by the use of a *fourth* satellite. This fourth satellite supplies a redundant measurement, since only three spheres of position are needed to make a fix. Hence, if our GPS receiver clock is accurate, the calculated fourth position sphere should intersect the three others at the same point. If it does not, it means that our GPS receiver clock is off, and our intelligent little gizmo will then correct its own clock to compensate for the difference. Thus our receiver is constantly verifying and adjusting its internal clock. As a result, not only does our GPS give us an extremely accurate fix, but it can also supply the time as precisely as an atomic clock! This is actually put to good use by people in need of very precise time (e.g. astronomers, seismologists) who use a GPS not as a position finding device but only as a clock.

182. Why do VHF radios have such a short range?

The "long waves" and "medium waves" used for AM (amplitude modulation) radio broadcasting have very long ranges, and so do "short waves." But VHF (very high frequency) waves have a very limited range. The reason for this is that VHF waves do not travel along the curved surface of the Earth the way other, longer waves do. They travel only in straight lines. This is also true of light waves and of the short radio waves used in FM (frequency modulation) broadcasting and television, and is the reason why so many communication towers are needed to cover extended areas.

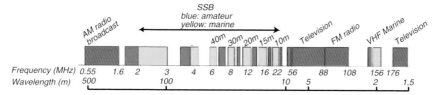

The radio bands used in communication. Only the medium and long waves used in AM radio broadcasting can go around obstacles, thanks to diffraction.

Actually, all electromagnetic waves travel in straight lines. Medium and long radio waves are an exception in that they can follow the curved perimeter of Earth and go around mountains thanks to the phenomenon of *diffraction*. Diffraction is the change of direction that a wave undergoes as it skims past the edge of an object in its path. This is also the phenomenon that causes ocean waves to go around a jetty.

All waves diffract as they skim by an object, no matter what their nature (sound waves, ocean waves, electromagnetic waves) or their wavelength. But diffraction is only important when the wavelength of a wave is greater than the size of the object.

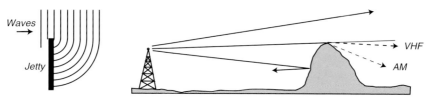

Just as ocean waves can move in behind a jetty by diffraction, radio waves can diffract behind obstacles. But the effect is small for shorter wavelengths, such as VHF.

Medium and long radio waves with wavelengths between 200 and 2000 meters diffract behind mountains and over the curved surface of Earth, traveling far beyond the horizons of their transmitters. VHF waves, which have a wavelength of about 2 meters, are essentially blocked by obstacles: they diffract very little around their edges and hence cannot "bend" to travel along the curved surface of the Earth. In order to transmit these waves, there has to be a direct line of sight between the transmitter and the receiver.

At sea, the range of a VHF wave is thus limited by the curvature of the Earth, and is given by the same formula as that for geographic visibility:

$$\text{Range in nautical miles} = 1.144\left(\sqrt{H_t} + \sqrt{H_r}\right)$$

where H_t is the height of the transmitting antenna in meters, and H_r is the height of the receiving antenna (with the effective range being about 20% greater than direct line of sight, however, thanks to atmospheric refraction).

For example, two boats with 40-foot masts would only be able to exchange radio messages in VHF up to a maximum separation of approximately 18 NM, whereas you can hear transmitters that are located high up, like Gibraltar (altitude 1400 ft), as much as 50 miles away.

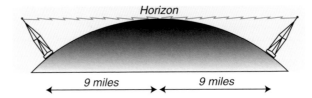

The short waves of the SSB radios,[22] which have a wavelength of between 10 and 40 m, also travel in straight lines, but they benefit from a particularity: they are reflected by the layer of the atmosphere called the ionosphere which is above 100 km altitude. When this layer is well established, multiple reflections become possible as the waves bounce back and forth between the ionosphere and the ground. Communication can then take place over thousands of miles.

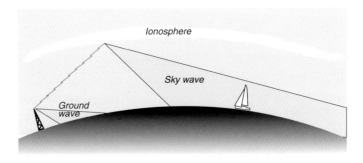

---

[22]SSB (Single SideBand) is the name given to short wave transmitters/receivers used at sea and by amateurs (ham) radio operators for long distance communication.

183. Why is radar unable to detect a boat with a fiberglass hull?

A fiberglass hull is as visible in daylight as any steel or aluminum hull; light rays bounce (reflect) right off it, some in our direction, and create the image of the boat on our retinas. But where radar waves are concerned, although metal is easy to detect, fiberglass is essentially transparent, "invisible." Why the difference?

Radar works with microwaves, the same kind as those used in microwave ovens. In these ovens, as everyone knows, food can be cooked in plastic bowls but not in metal pots. This is because microwaves, which are electromagnetic waves just as light is, behave more like radio waves than like light waves.

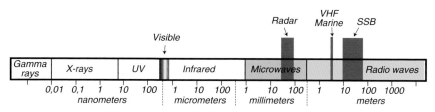

The spectrum of electromagnetic waves.

Radio waves, like all electromagnetic waves, consist of an electric field and a magnetic field moving through space (the two fields are perpendicular to each other and to the direction of propagation of the wave). When a radio wave encounters a conducting surface (i.e., made of metal), the electric field is "short-circuited," meaning that the wave's progress is blocked and bounces off in a new direction, as if reflected by a mirror.[23]

If the object encountered is not a conductor, however, the radio waves pass right through it. That explains how a radio or cell phone can work inside a building, behind walls, as long as the walls are not too thick and are non metallic.

Light waves, which are also electromagnetic, react the same way when they reach a conducting surface: a mirror reflects them because

---

[23] A brief explanation of the phenomenon is as follows. In a solid conductor, the electrons (enveloping the nucleus in layers called *valence shells*) can normally move around freely. When the incoming electromagnetic wave, called the incident wave, reaches the metal conductor, it excites the electrons in the conductor and makes them vibrate in the direction of the electric field of the incident wave. The vibrations generate a new electromagnetic wave of the same frequency as that of the incident wave, but opposite in phase. This new wave interferes with the incident wave, canceling it beyond its point of arrival on the metal surface, and adding to it "upstream" to form the reflected wave.

of the thin layer of silver behind the glass. But light waves do not usually pass through non-metallic materials as radio waves do. That is because of light's greater energy. You may recall that, according to quantum mechanics, all electromagnetic waves possess the mysterious property of being both a wave and a particle, the particle called a photon. The energy of a photon is inversely proportional to its wavelength: the shorter the wavelength, the greater the energy. Radio photons have very little energy, to the point where radio waves manifest themselves above all as long, "lazy" waves, not tight, compact particles. At the other end of the spectrum, gamma rays are very energetic and manifest themselves mostly as particles. Light is in between the two and manifests itself under both forms. It is energetic enough to make the electrons in atoms of most solid, non conducting materials vibrate. Its photons are first captured by the atoms, then almost immediately released, but in random directions: we say that the incident light is "scattered" (Q. 66). That is what makes it possible for us to see objects that, though not luminous themselves, are lit by the Sun or by artificial lights. The light that strikes them bounces off, scattered in all different directions, and some of the photons strike the retina of an observer.

In certain non-conducting materials, however, the electrons are bound so tightly to the atom's nucleus that the light energy is not strong enough to dislodge them and make them vibrate: instead of being scattered, then, most of the light passes right through the material. These are the "transparent" materials, glass for example.

And fiberglass boat hulls are to radar waves as glass is to light waves.

184. At what range can radar detect a small boat?

The answer to this question depends on the type of radar, the weather conditions, and the nature of the reflecting body.

The radar reflecting power of an object, referred to by the acronym RCS (Radar Cross Section), is expressed as the apparent surface of a metal sphere that would furnish the same echo. A sphere rather than a plane surface was chosen because a sphere always furnishes an echo, no matter how it is turned.

The radars used at sea work on two frequency bands: X band at a frequency of 9.4 GHz (gigahertz), or a wavelength of 3.2 cm,

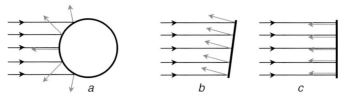

A sphere furnishes an echo no matter what its orientation (a), while a plane surface only furnishes one if it is perpendicular to the incoming wave (c). When this happens, the echo can be enormous, up to 1000 times stronger than the echo of a sphere with the same apparent surface.

and S band, with a frequency of 3 GHz, or a wavelength of 10 cm.[24] The use of the letters X and S goes back to the military origins of radar: they were the code names the British used during World War II. Recreational boats normally use only X-band radar, whereas commercial ships are usually equipped with both types of radar.

The X band allows smaller things to be detected but is more sensitive to rain and waves, which can muddle the image on the screen. The S band has a greater range and is less vulnerable to image distortion by rain and waves, but has poorer resolution: it provides fewer details. In general, commercial ships use X band near the coast and, when out at sea, they use S band with the range typically set for 24 nautical miles.

Some typical RCS values (in X band) are shown at right. In X band, the limit of detection in calm seas is 2.5 m², but, if one is to be easily visible, it is best if the RCS is at least 10 m². The sensitivity of the system decreases as the square of the frequency. That

Target	RCS (m²)
Commercial airliner	125
Small civilian aircraft	1.5
Automobile	10
Cargo ship	15 000
Motor boat (50 feet)	10
Sailboat (35 feet)	2
Rowboat, no outboard	0.02

means that, in S band, detection will be 10 times poorer.

So the answer to the question is this: in practice, a sailboat can be detected at a distance of 3 to 6 nautical miles. Radar reflectors have an RCS of about 2.5 m² in the best cases. That increases detectability a little, which is better than nothing, but the effect is marginal, particularly out in open seas where the big commercial ships use S band. In the worst case, 3 miles, that leaves only about 10 minutes to grasp the situation, perhaps communicate with the detected ship, and react. Not much time! As recreational boaters, if we have our

---

[24]The formula for transforming frequency into wavelength is: $\lambda = 300/\nu$, where λ is the wavelength in meters and ν is the frequency in MHz (megahertz). Note also that 1 GHz = 1000 MHz.

own radars (X band) aboard, it is much better to use them to detect other vessels than to count on others to detect us. As a matter of fact, if cargo ships can only detect us at distances of 3 to 6 miles, their RCS's are so enormous that we, with our modest radars, can detect them 12 miles off and more (and visually usually at 8 miles). On our small boats, we are better equipped (and certainly better motivated) than they to avoid a collision!

185. Are pirates still really a danger?

Piracy has always gone hand in hand with maritime commerce. A ship loaded with merchandise is slow and defenseless, an easy prey. But it does seem that the problem came close to disappearing from seas worldwide around the middle of the 19th century, thanks to the dissuasion of western naval forces. People had even begun to forget that piracy is a crime, often vicious and deadly. It had become an entertainment for us, a subject of fun, ships with billowing sails, skull and crossbones flags, dashing heroes capturing fair damsels...

But piracy has not disappeared; attacks on merchant ships reappeared as a not-so-rare phenomenon during the 1970's and have continued to increase. And if there are still a few pirates who climb up anchor chains with a knife clenched between their teeth, most attacks are perpetrated by violent armed bands equipped with hi-tech weapons.

There are hundreds of attacks each year. In 2002 there were 294 attacks, with 6 dead and 50 wounded. About 40 men that had been thrown overboard

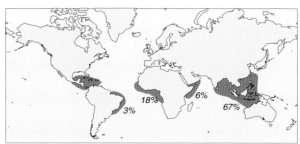

Pirate attacks worldwide in 2002.

were saved, but 38 others were reported lost. Most of these attacks were carried out on ships in ports or at anchor, but 18% took place on the high seas. Eleven ships were hijacked and 7 others disappeared completely [34]. Most of these attacks took place in the Far East (67%), particularly in Indonesian waters, but the Caribbean and the coasts off Africa and South America are also affected.

186. What causes *clapotis*, those choppy little standing waves often seen in front of breakwaters?

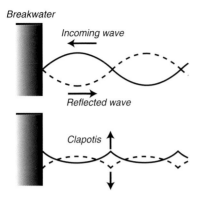

Breakwater
Incoming wave
Reflected wave
Clapotis

Clapotis is rather frequent in front of a quay or breakwater with vertical walls. The incoming waves are reflected by the wall and their direction is reversed (top figure). But since the reflected waves have amplitudes opposite to those of the incident (incoming) waves, the two waves simply cancel each other out and what is left is a relatively flat sea. The clapotis is what remains after the imperfect cancellation of the incoming and reflected waves. It occurs because the period and direction of incident waves is always a little variable (bottom figure).

That might seem to be an ideal way to calm the sea at the entry to ports, and many breakwaters have been constructed using the principle of cancellation of waves by reflection [85]. Unfortunately, during big storms the stresses on the walls are so great that they can be swept away. Current practice is to *absorb* the incoming waves rather than try to *reflect* them (Q. 187).

187. What are those odd-shaped cement things you sometimes see piled up in front of breakwaters?

Those curious-looking objects are *tetrapods*, a word meaning "four-footed." They are designed to absorb the energy of waves striking a breakwater. When piled up, they provide a great many small spaces and chinks

into which the water can rush and "exhaust its energy," bouncing around and rubbing against the rough inner surfaces.

Their specific shape lets them interlock easily, preventing them from being dislodged during storms while maximizing the absorption of wave energy. Ordinary rocks are often used for the same purpose. That solution is less expensive but also less efficient.

188. Why is there a risk of collision if the bearing of an approaching ship doesn't change?

Do you by any chance remember Thales' theorem? "Equidistant parallel lines describe segments of equal lengths on any intersecting straight line." So, if two boats, A et B, that maintain both their speed and heading are going to collide, the lines connecting them at given time intervals are parallel. Hence, if boat A sees boat B at a constant bearing, (for example, if boat B is always right behind a shroud), there is a serious danger of collision.

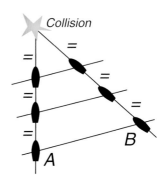

189. How far away can a flash from a signal mirror be seen?

You always find a signal mirror tucked into a packet of emergency equipment. Are such mirrors really useful items or are they just fit for playing boy scout games?

Actually, those little mirrors can be seen from quite a distance. If you project the image of the Sun as far as you can, you lose it after about 150 feet. But an observer much farther away would be able to see the reflection quite clearly [49]: with a 4 x 5 inch mirror, the reflection is visible 10 nautical miles away.

You have to aim the beam carefully, however. One good way is to aim through the light circle in the middle of the mirror, direct the reflected beam through the V made by two fingers of your outstretched hand, and use this to help send the beam towards its target.

190. Why were the letters "S O S" chosen for the distress signal?

Did anyone ever tell you that S O S stood for "Save our souls"? Pure myth! The letters were adopted by an international commission in 1906 because, in Morse code, those three letters translate into a sound signal that is particularly easy to recognize (· · · — — — · · ·). The other call for help, "May Day," which is used in voice communications and was adopted internationally in 1948, comes from the French, "m'aidez" (help me).

191. Exactly where was Christopher Columbus's landfall on his first voyage?

On October 12, 1492, Christopher Columbus landed at Guanahani, a small island on the Atlantic side of the Bahamas. Guanahani was the name the inhabitants had given their island, but Columbus rebaptized it *San Salvador* (Holy Savior). So where exactly is Guanahani? The experts have been debating that question for centuries [51].

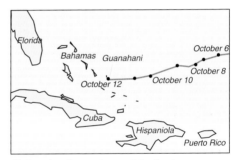

Columbus's route on his first voyage.

An island called San Salvador certainly exists today, but that is just the new name adopted by Watling Island in 1925. Even though this island is a plausible candidate for the historical landing site, it was renamed San Salvador mainly in the hopes of attracting a few tourists, not because the experts had come to the conclusion that it was *the* island.

Whether because Columbus himself made errors (perhaps the result of his enthusiasm at finally reaching land), or because of errors made by copyists (Columbus's personal log is only known in the form of an incomplete copy made by Bartolomé de Las Casas), there is no island that clearly corresponds to the description of Guanahini in Columbus's log.

To celebrate the five hundredth anniversary of the "discovery of the New World," the American *National Geographic Society* had a new translation made of Columbus's log and, aided by computer modeling, did a detailed recreation of his described route from island to island, all the way to Cuba. The conclusion of this study was that

Columbus landed first at Samana Cay, a little island about 10 miles long, uninhabited and hard to access [36].

The proofs are not completely convincing, however. Although the route Columbus took to explore the neighboring islands seems to coincide relatively well with a departure from Samana Cay, the geography of that island does not match the description he gave of it. Guanahani had a protected bay and a lagoon, neither of which can be found at Samana Cay. From that viewpoint, Watling Island is a better fit. And besides, Gua-

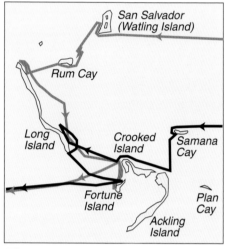

The two most probable landfall sites.

nahani was inhabited by "Indians," as Columbus called them, believing himself to be in the Far East. The original populations have been exterminated since, but there are many archaeological remains on Watling, while Samana Cay doesn't even have a source of fresh water [15].

Although it seems pretty clear that the choice could only be between Watling Island (the current San Salvador) and Samana Cay, the identity of the mythic landfall has yet to be resolved.

192. Who was the greatest navigator of all time?

So many names come to mind: Pytheas, Columbus, Magellan, Vasco da Gama, Bougainville, etc., and more recently, Slocum, Chichester, Tabarly, Knox-Johnston... This is a delicate matter, with chauvinism and bias hard to eliminate, but we will venture an opinion that at least tries to be objective. If by navigator we mean a man who had it all, a profound understanding of the sea, a great talent for running a ship, commanding a crew, and the difficult art of exploring unknown coasts, and who, finally, traveled over all the seas of the world, one name stands out above all the others: James Cook.

It was the Age of Enlightenment, a century where reason and knowledge were prized. Cook's voyages of exploration, financed by the Academy of Sciences and the British Admiralty, were motivated by relatively unselfish goals, free of thoughts of conquest, unlike the explorations of preceding centuries. His ships carried scientists and

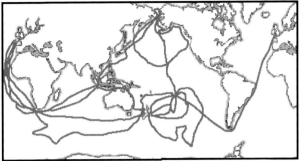

Captain James Cook (1728-1779) — His Pacific voyages of exploration.

artists who would bring back faithful images of their voyages and discoveries.

During his three epic voyages around the world, Cook put the Pacific, which until then had been practically unknown, "on the charts" [32]. Covering 150,000 nautical miles, from Australia to Siberia, from Oregon to Tahiti, his voyages of exploration had an impact that can still be felt today.

193. Where is Robinson Crusoe's island?

Robinson Crusoe is the most famous shipwrecked sailor in history. His adventures, as told by Daniel Defoe in his narrative published in 1719 and re-edited many, many times since, have captivated countless children and adults. It is one of the written works that has been one of the most translated and published, worldwide, along with the *Bible*, the *Iliad* and the *Odyssey*, and Euclid's *Elements of Geometry*.

In his journal, Robinson tells us that his island is located just off the delta of the Orinoco River at latitude 9° 22′ N. He also tells us that the large island he can see to the north-west is Trinidad. As for Friday, he is a Carib from what is now Venezuela.

That island does not exist, no more than Robinson Crusoe himself ever did. Robinson and his island were simply products of Defoe's imagination. There is, however, a *Robinson Crusoe Island* in the Juan Fernandez archipelago about 400 miles off the coast of Valparaiso, that the Chilean government rebaptized as such in the 1960's, hoping to attract tourists. There, you can visit "Robinson's cave" and his terrace with the panoramic view. And "a Robinson" actually did live for four years on this formerly uninhabited island. The man's name was Alexander Selkirk.

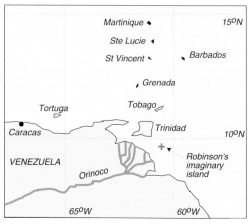

Robinson Crusoe's imaginary island is located off the delta of the Orinoco River.

Selkirk was a Scot who had sailed as quartermaster with an English privateer pursuing Spanish ships. When it turned out that he and his captain could not get along, he asked to be set ashore on this island, where he lived alone until picked up in 1709 by another English privateering ship, the DUKE, on which the famous former pirate, William Dampier, was sailing. Two years later, Selkirk arrived in England, where his strange adventure became a sensation. His memoirs were published in a magazine, and Daniel Defoe drew his inspiration from them to create his legendary character [66].

194. Who were the first navigators to sail solo around the world?

> " The voyage which I am now to narrate was a natural outcome not only of my love of adventure, but of my lifelong experience..."

These are the opening words of the memoirs of Joshua Slocum, the first solo circumnavigator of the globe. In 1895, at the age of 51, after having spent most of his life at sea, he set out on a little sailboat that he had rebuilt himself, the SPRAY.

He had few emulators in the following decades, still no more than a dozen of them by 1975 [31]. That was the era of true adventurers out to discover a world that could be discovered in no other way, and, for most of them, to search for meaning in their lives. None of

The SPRAY *and Joshua Slocum around 1898.*

them were in a hurry, making stopovers, sometimes for weeks, wherever they felt like it. The books they wrote inspired a generation of young would-be navigators who dreamed of experiencing that exotic, sublime, challenging world that was on the verge of disappearing.

Since then, plane travel has made long distance travel banal, and solo circumnavigations seem mainly to attract sailing professionals seeking big media coverage for their well publicized exploits.

The single-handed circumnavigators through 1975

Navigator	Nation-ality	Boat	Length	Dates	Remarks
Joshua Slocum	USA	SPRAY	36 ft	1895-98	first solo
Harry Pidgeon	USA	ISLANDER	33 ft	1921-25	via Panama
Alain Gerbault	F	FIRECREST	36 ft	1923-29	
Louis Bernicot	F	ANAHITA	39 ft	1936-38	
Vito Dumas	Arg	LEGH II	31 ft	1942-43	
Jean Gau	F	ATOM	30 ft	1953-57	
Marcel Bardiaux	F	LES 4 VENTS	31 ft	1950-58	first E-W
Pierre Auboiroux	F	NÉO-VENT	27 ft	1964-66	
Francis Chichester	GB	GIPSY-MOTH	53 ft	1966-67	one stop
R. Knox-Johnston	GB	SUHAILI	32 ft	1968-69	no stops
Bernard Moitessier	F	JOSHUA	39 ft	1968-69	no stops
Chay Blyth	GB	BRIT. STEEL	60 ft	1970-71	E-W, no stops
Alain Colas	F	MANUREVA	68 ft	1973-74	trimaran

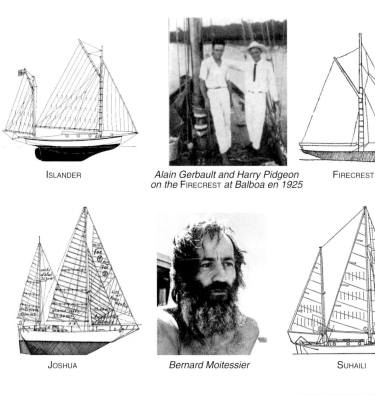

ISLANDER

Alain Gerbault and Harry Pidgeon on the FIRECREST *at Balboa en 1925*

FIRECREST

JOSHUA

Bernard Moitessier

SUHAILI

Sir Francis Chichester
(reproduced with authorization
of the Estate of Francis Chichester)

GYPSY MOTH IV

*Robin Knox-Johnston with
provisions for 10 months.*

A few famous solo circumnavigators and their boats.

195. Where are the most beautiful cruising spots in the world?

The world is not what it used to be. In Slocum, Stevenson or Gerbault's time, its most distant shores were really only accessible by boat, either privately owned or chartered. Today, rare indeed is the enchanted hideaway you cannot reach by jet plane plus taxi boat.

Still, there are a few remaining shores, ports, and anchorages that can really be best appreciated if you arrive on your own boat. Everyone cherishes a secret paradise they would rather not reveal. Let us therefore consult, with gratitude, a list that Gilles Martin-Raget has made public [47], and to which we have added a few of our own favorites. Summarizing:

- In the U.S.: New England, Puget Sound, Gulf of Florida.
- On Europe's Atlantic and North Sea coasts: Cowes and The Solent, Ireland, Scotland, Sweden, Brittany, La Rochelle and offshore islands.
- In the Mediterranean: Corsica, Sardinia, the Balearic islands, Croatia, Greece and Greek islands, Turkey.
- The "tropical paradises:" the Caribbean, the Bahamas, the Seychelles, the Maldives, Polynesia.
- The end of the world: Australia, New Zealand, Tierra del Fuego.

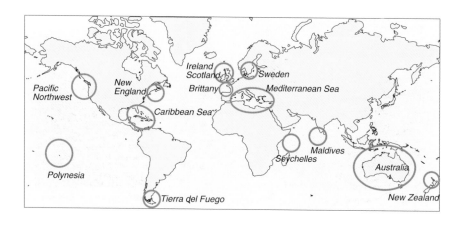

Life Aboard

196. Why do some people get seasick?

Were it not for certain control centers in our brains that maintain our equilibrium, we would topple over like rag dolls. To determine the exact position of up, down, sideways and points in between, the brain takes the information relayed to it by our senses, calculates our orientation in space, and informs our muscles how to react so as to keep us upright.

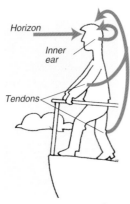

On a boat, our brain receives contradictory information.

The brain gets most of its information from our inner ear, which acts as a sort of carpenter's level, but it also uses data transmitted by our eyes, muscles and tendons. On terra firma, all goes well because the different sources of information are in agreement. But at sea, things are not so clear because the movement of the boat mixes up the signals that our brain has to interpret. What our inner ear senses no longer agrees with what our eyes see or our muscles feel. We are like an airplane pilot whose eyes tell him one thing while his instruments tell him another. Which to trust? Is he "seeing things" or are his instruments malfunctioning? He cannot tell. He has to decide, hesitates, his palms become sweaty...

This is the situation we face at sea. Receiving contradictory information from our senses, the brain may immediately decide to disregard some of the information, in which case we will feel fine. But if confusion continues to reign, nervous tension builds up and results in seasickness. Within a few hours, if the brain has still not adapted to the situation, vomiting occurs. This often relieves the nervous tension, hence the temporary relief one can feel then.

Vision plays a big part in the way our brains judge orientation. When we are on deck, the movement of the boat sensed by our muscles and inner ears is confirmed by the inclination of the horizon. If we go below, though, our eyes will only be able to judge the boat's motion by what they see in the cabin, while our inner ears continue to sense the boat moving in the waves. This kind of conflict can easily lead to seasickness. It is best to be out on deck. But you mustn't lie motionless in the cockpit; you have to "ride the waves" like a horse-back rider going over jumps. That means looking around at the sea so that your brain can anticipate what posture to adopt. It is also a good idea to get up and walk around, if possible. It helps develop "sea legs" [54].

197. Why is it best to use red light in the cabin at night?

During your night watch, you might decide to go below to check a chart or heat up a cup of soup. If you switch on the normal cabin lighting, it will take you 20 minutes or more to recover your night vision when you go back on deck. This is the same phenomenon that temporarily almost blinds you when you enter a dimly lit room after being out in bright sunlight.

The retina of the eye contains two types of light-sensitive cells: *rods*, which are sensitive to low levels of light but cannot distinguish color, and *cones*, which are sensitive to fine detail and color but only in good light.[1]

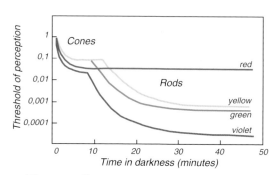

Color sensitivity of cones (left of break) and rods (right of break) as a function of time spent in darkness. Rods are insensitive to red and most sensitive to violet.

The excellent sensitivity of our rods in low light is due to the presence of a substance called "visual purple[2]." Unfortunately, this substance is destroyed by bright light and it takes 20 to 30 minutes of darkness for it to be reconstituted. Now, since visual purple is itself

---

[1]Three different types of cone cells, each type sensitive to a different range of wavelengths of light, work together to produce color vision. But since all rod cells are sensitive to the same range of wavelengths, they cannot produce the sensation of different colors.

[2]This term is misleading, visual purple is really a shade of red — the scientific name is *rhodopsin*.

reddish in color, it absorbs the light from every color except red and is actually quite insensitive to red light at moderate light levels.

And so, when we need to see without compromising our night vision, the solution is to use red light. Our cones can function in it well enough to let us read a map or pour out a cup of hot water without destroying the visual purple in our rods.

This is why red light is generally used on the compass and navigational instruments, and also why it is a good idea to have red lights in the cabin and at the chart table. The one problem with this solution is that our eyes cannot distinguish colors well in red light and, as a result, the magenta used on charts to indicate buoys is hard to see. For this reason, airplane pilots sometimes prefer using green light to examine their colored charts. Our eyes are much more sensitive to green than to red and pilots can see the necessary details without using too strong a light and completely losing their night vision.

At sea, then, if normal levels of red light do not suffice, use the lowest level of white light that works, for example by using a flashlight covered with a handkerchief, to minimize the loss of night vision.

Would eating large amounts of carrots work miracles for someone's night vision? Well, it is true that our rods synthesize visual purple using *carotene*, a substance that carrots are full of. But carotene is also found in other vegetables and fruits as well as in the fat of plant-eating animals. A normal diet amply furnishes all that we need of it and, unless someone has a deficiency of vitamin A (the vitamin needed for the synthesis to occur), our night vision would not improve no matter how many more fruits and vegetables we ate.

198. What is the origin of the term "cockpit"?

Cockfight in a "cockpit."

It might seem obvious that the term "cockpit," referring to the depressed aft area of a small boat from which the vessel is steered, is a direct reference to the small arena where the blood sport of cockfighting takes place, to the delight of excited fans. After all, a boat's cockpit does vaguely resemble a fighting cock's cockpit, at least in size and shape. But, as it turns out, the story is probably more tangled than that.

By the 16th century, from its original association with the cock-fighting arena, the term "cockpit" had expanded to mean any small-ish or confined place of entertainment or frenzied activity. Shake-speare used it in that sense in *Henry V* to refer to the area around the stage of a theater. Meanwhile, keeping its association with vi-olence and bloodletting, it also came into the lexicon of the Royal Navy of the 17th and 18th centuries, when it referred to the area on military ships where junior officers were stationed and where, during naval battles, the wounded were often brought to be attended to — a bloody and violent business then.

So did the term "cockpit" just mutate from there, move from inside a warship to outside on a small vessel, lose its connotation of bloody violence while keeping the idea of a small, confined space, and meanwhile pick up a new association with steering? What could explain such a transformation?

Before attempting to answer that, we should take a look at the word "coxswain," once long ago written "cokswayne," meaning the servant (swain) of the cockboat (think cockleshell).[3] This was the ship's sailor who was in charge of a small boat and its crew, and who usually steered it. And while the coxswain steered his cockboat, he was sitting (or at least had his feet in) a little depression — a "pit."

All right, then, which is the true origin of the "cockpit" that entered the yachting world in the 19th century? Was the cock in question a gamecock or a cockboat?

It seems likely that both references were involved — by analogy, association, fusion or contamination. The Royal Navy's term "cock-pit" was already well known and had some of the required meaning. And the steering stations of small, open cockboats were very like those of the new pleasure boats, but seem not to have had a specific name. It would have been serendipity, a match waiting to be made. And once made, of course, it was an easy step to extend the use of the term to cover the small steering compartments of airplanes and racing cars, too.

---

[3]Indeed, the cock- of "cockboat" and "cockleshell" can itself be traced back through the French *coque*, meaning "hull of a boat," to the Greek word for conch, a sea creature with a fine, protective "hull."

199. Why does a dinghy slide backward as we try to climb out?

There you are, about to go ashore, your dinghy has just nudged the wall of the pier. Full of confidence, you stand up and lean forward to slip the line through the iron ring or simply try to step out and *voilà*, the dinghy slips away under you.

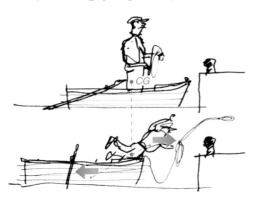

When you lean forward in your dinghy to tie it up, it slips backward under you.

This disconcerting little trick it plays is due to what physicists call "the conservation of linear momentum." If one part of a free body moves in one direction, the rest of the body moves in the other direction in such a way that the global center of gravity remains in the same place.

When we walk, friction between the ground and the foot that touches it prevents our body from sliding backward as we swing our other foot forward. This friction is precisely what makes walking possible. Now, our dinghy slips through the water with very little friction. When we change our position in it, our common center of gravity (we + dinghy) must remain in the same place. If we move towards the bow, since friction between the dinghy and the water does not counteract the movement, the dinghy moves back so that the center of gravity stays in the same place. Our floor slides backwards under us, we lose our balance and...plop!

200. Can a mast serve as a lightning rod?

Being caught in a storm at sea is a terrifying experience. Especially at night. Even more so if one thinks that the mast, like a lightning rod, just seems to be designed to attract lightning. And the risk is real. Along the coast of Florida, for example, where storms are particularly numerous and violent, lightning strikes 5% of the boats annually (at sea or in marinas), and it is estimated that *every* boat there will be hit at least once during its lifetime [75].

Surprisingly enough, lightning strikes seldom cause serious damage, usually just tiny holes or cracks in the hull, too small to cause leaks.[4] The greatest risk is for the crew and the electronic equipment. Is it possible to reduce this risk?

It is true that a *metallic* mast "attracts" lightning. Like a lightning rod atop a building, it is tall and thin and a good conductor of the positive electric charges that accumulate at the surface of the Earth (Q. 101).

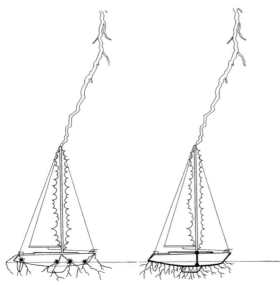

Some people think that insulating the mast from the rest of the boat and from the water will minimize this attraction. Not so. Even insulated, the electrical resistance between the base of the mast and the water is not enough to block the upward flow of positive charges at the water's surface and avoid a strike. Lightning easily

Path of lightning in a boat with insulated mast (left) and with mast grounded to hull (right).

makes its way through several miles of air, and air is a very good insulator. A few extra feet of it would hardly pose much of an obstacle!

It is better to channel lightning along a predetermined path so that it does not choose its own path through the hull and the crew, in essence by installing a lightning rod.[5] To accomplish this, all one has to do is ground the mast (and, if possible, the stays and shrouds) by attaching it to a large metallic mass in contact with the water, such as the propeller or the keel.[6]

---

[4]Only in 5% of cases is damage significant: lightning can put a 1/8 inch or even larger hole through a hull — a risk that does not exist for metal hulls.

[5]Since our bodies contain a lot of water, our electrical resistance is much lower than that of air.

[6]Even if painted or sheathed in fiberglass, a keel is a good ground, but the cables connecting it to the mast must be heavy guage and free of kinks.

201. Taste aside, is there any particular reason to eat hot, spicy food in the tropics?

Many countries with hot climates have spicy cuisines,[7] and for a long time no one knew why. It was speculated that eating hot, spicy food made people sweat and so helped them feel cooler, or that the spices were used to disguise the taste of spoiled food, or even that a great many spices just happen to grow well in hot climates but not in more northerly ones.

Recent studies have provided a likely explanation: spices have an antibiotic effect [69]. In hot climates where food, especially meat, goes bad quickly, onions, garlic[8] and hot peppers prevent (or at least slow down) the development of most bacteria that cause spoilage. That does seem reasonable: as temperature rises, the activity of bacteria and fungus often increases. Certain plants in hot climates have thus "learned" to protect themselves from attack by producing antimicrobial substances. We get the benefit of those substances, too, when we eat spicy food. So when we are visiting a hot country where conditions for storing and refrigerating food leave something to be desired, it is probably a good idea to do as the Romans do... and eat as the locals eat.

202. Is scurvy still a risk today on long cruises?

In the past, scurvy caused the death of many a sailor. Vasco da Gama lost two thirds of his crew to the disease during his voyage to India via the Cape of Good Hope in 1499, and Magellan lost 80% of his crew to it while crossing the Pacific in 1520. In the 17th and 18th centuries when long ocean voyages had become common, it killed 5000 seamen every year.

We now know that scurvy is caused by a deficiency of vitamin C, for which the chemical name is ascorbic acid (the word "ascorbic" actually means "no scurvy"). The role of vitamin C is to help produce collagen, an important protein in the connective tissue of teeth and bones, skin and tendons. When we cut ourselves, our body produces collagen to close up the wound.

---

[7]This is true of India, South-East Asia, Mexico and the Caribbean, countries around the Mediterranean, and certain others, but not of the Pacific islands nor of many tropical African countries, where spices were not introduced.

[8]Onions probably originated in West or Central Asia and garlic in Central Asia.

Scurvy appears when one has had a vitamin C deficiency for 5 to 6 months. It is a terrible disease. Initially, the gums bleed and the teeth grow loose in their sockets, making it hard to eat; small cuts in the skin no longer heal; the tendons become swollen and painful. At advanced stages the tissues of the body putrefy, the skin blackens, and death follows.

Vitamin C is found in citrus fruit (lemons, oranges, grapefruit, limes) and certain vegetables,[9] but cooking destroys it and it decomposes naturally when exposed to air.

The Dutch discovered the benefits of citrus fruit back in the 17th century, and orange and lemon juice was promptly put on the menu for their crews.[10] They had also discovered that sauerkraut kept scurvy at bay, and served their crews a pound of it per man twice a week.[11] But it was not until 1795 that the British Admiralty began furnishing its men with lemon juice. Until then, scurvy claimed more victims in the English Navy than battles did!

Today, scurvy is no longer a problem since it is rare for anyone to remain at sea for over 6 months without calling at a port. We need about 50 mg of vitamin C daily and one lemon or orange per day can furnish that. But on long passages, if the fresh fruits and vegetables run out, it is a good idea to take vitamin C in pill form.

203. Why does soap work so poorly in sea water?

Let's begin by taking a look at how soap works in fresh water. Soap is a mixture of a fatty (greasy, oily) substance and an alkaline substance such as lye. The fatty substances in soaps are made up of long chains of carbon and hydrogen atoms. At one end of each chain is a group of atoms that is attracted to water (hydrophilic), while the other end is repelled by water (hydrophobic) but is attracted to other fats.

Getting rid of non-greasy dirt on an item simply requires rinsing in water; no soap is necessary. But greasy dirt remains stuck on

---

[9]Melons, broccoli, red and green peppers, tomatoes and lettuce are good sources of vitamin C.

[10]This discovery was made by accident. In 1600, out of four ships that sailed for India, one had lemons on board and the three others did not. Most of the sailors on the ships without lemons fell ill while very few on the ship with them did.

[11]Sauerkraut, which is fermented cabbage, contains 10 to 15 mg of vitamin C per 100 grams.

the item because grease does not dissolve in water. Now add some soap. The lye in it immediately dissolves, liberating the hydrocarbon chains, whose hydrophobic ends stick to the greasy dirt, while the hydrophilic ends are attracted to the water molecules. If we agitate the water or rub the item being washed, the chains of hydrocarbons lift up and detach the bits of grease to which they have attached themselves. In a sense, the grease has been dissolved.

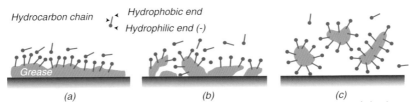

The hydrocarbon chains in soap attach themselves to the grease (a), dislodge it by agitation (b), and carry it away (c).

Unfortunately, soap washes poorly in hard water and seawater. The heavy concentrations of mineral salts produce positively charged ions in solution (ions of sodium, calcium, magnesium, iron, etc.) which are attracted to the hydrophilic (negatively charged) ends of the hydrocarbons, preventing them from attaching to the greasy dirt.

Detergents, on the other hand, work relatively well in salt water because, at the hydrophilic ends of their hydrocarbon chains, they have synthetic compounds that are only moderately attracted to the ions of mineral salts.

Some detergents work better than others in salt water, however, and it is well known in the cruising community that Procter & Gamble's dish detergent *Joy*[12] does the best job, be it for hair, clothing or dishes. There are also liquid soaps specially formulated to work in salt water.

204. Can seawater be used for cooking?

Anyone who has ever tried cooking spaghetti in seawater knows the answer to this question: the results are almost impossible to swallow.

That's not too surprising when you stop to think about it. The recipe for spaghetti for three or four people is half a pound of pasta in about 3 quarts of water with a teaspoon and a half of salt. At about 5 grams of salt per teaspoon, that gives 2.5 grams of salt per

---

[12]When questioned, *Procter & Gamble* declined to reveal their secret.

quart. But there are 37 grams of salt per quart in seawater — 14 times too much!

205. Why does a piece of fish spoil faster than a piece of meat?

No question about it, a dead fish starts to smell bad sooner and decomposes more quickly than meat does. A fish lasts scarcely a day without refrigeration. Several factors are involved [62].

Fish contains a special chemical substance, trimethylamine oxide, that helps it avoid dehydration in its salty environment (Q. 40). This oxide is found in meat, too, but in smaller quantities. Once a fish is dead and out of the water, bacteria attack the oxide and transform it into trimethylamine. This is the substance with the well known odor of spoiled fish.

A second factor is that the texture of fish differs from that of meat. We can easily crush a piece of fish between our fingers, whereas the flesh of a mammal or a bird is much more resistant. The muscles of the latter are stronger, denser, stringier, because they have to do so much work to overcome gravity, something not required of a fish floating in water. And since the flesh of fish is less dense and composed of smaller molecules, bacteria and enzymes can attack it more readily.

A final factor is that fish are cold blooded: their internal temperature is as cool as the water they swim in. Bacteria that the living fish normally harbors are adapted to this low temperature and begin multiplying rapidly when the fish, once dead, is brought out into warm air. The reverse is true for a warm blooded animal, whose internal population of bacteria are adapted to a higher temperature and become less active (at least for a while) when the animal's flesh cools down at death.

206. Although oxygen is heavier than nitrogen, these two gasses never form separate layers in air. Why then does butane or propane tend to separate out and accumulate in the bilge?

Perhaps you remember a lecture on Dalton's experiment from your high-school chemistry course: if you join the openings of two balloons, one filled with oxygen and the other with nitrogen, the contents of the two balloons inevitably mix. That is what happens in the atmosphere. Even though oxygen is 14% denser than nitrogen, the two gasses never stratify there. If they did, we would have pure oxygen at ground level and nitrogen for our upper atmosphere.

This is the case because the densities of these two gasses are nearly the same, so that the normal movements of their molecules overcome the gravity that affects them to the same degree. Also, the atmosphere itself is always in movement, constantly stirring the two gasses around.

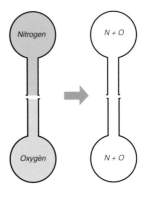

But this is not the case for heavier gasses in an enclosed space. There, kinetic energy due to the natural movement of heavy molecules is not strong enough to overcome gravity. The Cave of the Dog in Spain is a good example of this. Carbon dioxide (CO_2), which is 50% heavier than air, accumulates near the floor of the cave and has been famously responsible for asphyxiating several dogs. The same situation arises in a boat. If there is a gas leak, propane (C_3H_8) or butane (C_4H_{10}), which are one and a half and two times as heavy as air respectively, accumulates in the bilge and creates the dangerous situation we all know about.

207. With all the seawater available in case of fire on a boat, is it really necessary to have fire extinguishers aboard?

With water, water everywhere, it might indeed seem redundant to equip a boat with fire extinguishers. Why not just count on using a bucket of seawater?

Fire is a combustion, a chemical reaction, involving three components: a *combustible*, an *oxidant*, and a source of *heat*. The combustible is the material that catches fire, the oxidant (for ordinary fires) is the oxygen in the air, and heat is what causes a fire to break out and permits it to maintain itself.

These three elements are necessary for a fire to exist. Eliminate one of the three and the reaction stops. And so in order to put out a fire one can:

- reduce the quantity of oxidant (air) available to the combustible, for example by putting a lid on a skillet that has caught fire or throwing a blanket over a blaze or by using an extinguisher to cover one with a layer of foam or powder so as to insulate the combustible from the oxygen of the air,
- cool the combustible, for example by throwing water on it,
- or, obviously, eliminate the combustible, for example by turning off the gas in the case of a gas fire.

Lowering the temperature by throwing water on a fire is very effective for what are called "dry fires" (also called class A fires), where the combustible is solid: wood, paper, plastic, fabric. Not only does the water cool the combustible but it vaporizes to create an insulating layer, a vapor barrier, that prevents contact between the combustible and the oxygen of the air.

On the other hand, water should never be used on fires where the combustible is a liquid (called class B fires), such as grease, oil, paint, solvents, gas or diesel fuels. The temperatures of such fires are so high that water vaporizes instantly on contact with the combustible, making the flaming liquid spatter and spreading the fire rather than extinguishing it. Additionally in such cases, since water is the denser of the two liquids, the flaming combustible floats on the water, again tending to spread the fire further.

For these class B fires, the solution is to use a powder or foam extinguisher.

208. Why shouldn't plastic trash be discarded in the sea?

Space is at a premium on a boat and garbage can smell, too, so it became something of a sailor's tradition to toss waste overboard. In the past, the oceans could handle that well enough, but this is no longer the case. The composition of our trash has changed and quantities have increased considerably: it is estimated that commercial ships throw more than 6 million tons of waste overboard annually. The most dangerous kind of trash is plastic, that wonderful material we use so much of, but which is not, by nature, biodegradable.

Plastic items that are tossed overboard get caught in fishing nets and propellers, obstruct cooling pipes in engines, pollute beaches and kill countless fish, birds, and marine reptiles and mammals that swallow them or get tangled in them. A simple plastic sack looks like a jellyfish but is not digestible; any animal that mistakes one for the other may die of suffocation or of an intestinal blockage.

The interna-
tional convention
for the prevention
of marine pollution
from ships (MAR-
POL) forbids the
practice by any
vessel of throwing
plastic waste over-

Decomposition time for trash tossed overboard	
Paper	2 to 5 months
Orange peel	6 months
Milk carton	5 years
Tin can	10 to 100 years
Plastic bag	450 years
Aluminum can	200 to 500 years
Plastic bottle	100 to 1000 years
Glass bottle	4000 years

board anywhere at sea. Food, cardboard, metal (tin cans) and glass can only be discarded in the water when a boat is more than 12 nautical miles from any coast.

209. Why do sailors wear "bell-bottom" trousers?

Actually, they don't any more. At least American sailors don't, not since January 2001. On that date the U.S. Navy consigned the traditional denim dungarees with the 12-inch flare at the end of the pant leg to mothballs, replacing them with straight-legged trousers.

Although bell-bottoms had been in occasional use in the U.S. Navy since around 1810, the tradition officially began in 1817. In that year, the Navy adopted the bell-bottom design for the stated reasons that the flared cut permitted sailors to roll their pant legs up more easily when washing down the decks, and were also easier to take off quickly when it was necessary to abandon ship or when someone was washed overboard. The pant legs could also be knotted to trap air and serve as a life preserver.

Other navies, including the British Navy, also adopted the bell-bottom pant leg for the same practical reasons.

210. Why is our vision blurry underwater?

If you have ever tried to scrape your keel or clean off your propeller without wearing a mask, you not only had to put up with the salt water stinging your eyes, you found your vision badly blurred, too.

This is because our lenses, the optics of our eyes, are designed to function in air, not water. They work like magnifying glasses or photographic lenses, but since their index of refraction is nearly the same as that of water, once in the water they are essentially useless.

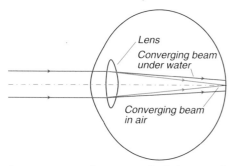

Lens
Converging beam under water
Converging beam in air

Light rays, instead of focusing on our retina, continue on nearly parallel paths. That makes us horribly farsighted. We could not even identify a coin held out at arm's length.

With a mask on, though, everything is fine. The cornea is in contact with air and the lens recovers its normal ability to focus.

211. For onboard reading, what are the greatest books about sailing and the sea?

Everyone will have their own preferences here, but there are a few literary works that have to be on anyone's list. Here is one attempt at such a list.

Non fiction

William Bligh, *A Narrative of the Mutiny on Board His Majesty's Ship Bounty*. The famous mutiny and some of its aftermath related by Captain Bligh himself. No matter what one thinks of the man, his incredible 3600-mile Pacific voyage on an open lifeboat as far as Timor is one of the greatest sea adventures of all times. (Also see *The Bounty* by Caroline Alexander.)

Alain Bombard, *The Voyage of the Hérétique*. In 1952 a medical doctor sails across the Atlantic on an inflatable raft with next to no provisions to demonstrate the possibility of surviving shipwrecks and plane crashes at sea.

Louis-Antoine de Bougainville, *A Voyage Round the World*. Bougainville's famous voyage of discovery and exploration in the South Pacific, 1766-1769, preceding Cook by several years. He presents Tahiti as an earthly paradise peopled by "noble savages," a concept that nourished the utopian thought of Jean-Jacques Rousseau in pre-revolutionary France. (See also *News from New Cythera; a report of Bougainville's voyage, 1766-1769* by L. Davis Hammond.)

Alvar Nunez Cabeza De Vaca, *Castaways*. An extraordinary true tale of Spanish explorers shipwrecked and marooned on the newly discovered North American continent, 1528-1536. More a land story than a sea story, but what a story!

John Caldwell, *Desperate Voyage*. Caldwell's harrowing account of finding himself stranded in Panama after World War II and setting out single-handed on a 9,000-mile journey aboard the 29-foot PAGAN to rejoin his wife in Sydney, Australia.

Francis Chichester, *Gypsy Moth Circles the World*. The solo round-the-world voyage of the famous navigator, then 65.

Christopher Columbus, *The Log of Christopher Columbus*. A translation/reconstruction of the logbook kept by Columbus himself during his first great voyage of discovery.

James Cook, *The Journals of Captain James Cook*. Eleven years of discoveries by the greatest explorer of all times. (See also Nicholas Thomas's book listed below.)

Jacques-Yves Cousteau, *The Silent World*. The book that first made the breathtaking world beneath the sea available to all of us.

Richard Henry Dana, *Two Years Before the Mast*. A Harvard law student who sailed as a crew member on a ship bound for California in 1834 writes about "the life of a common sailor at sea as it really is," showing the abuses to which his fellows were subject.

Charles Darwin, *The Voyage of the Beagle*. The sea voyage that changed the way we see ourselves and all living things.

Gabriel Garcia-Marquez, *The Story of a Shipwrecked Sailor*. A sailor spends ten days clinging to a life raft, as retold by the author of *One Hundred Years of Solitude*, then a young journalist.

Alain Gerbault, *Alone Across the Atlantic*. The narrative of one of the first solo circumnavigators (1923-29) who eventually settled in the islands of the South Pacific.

Sterling Hayden, *Wanderer*. Hayden walks away from a successful acting career and a marriage in tatters to set sail for the South Seas, taking along his four young children, the objects of a bitter custody battle. An unforgettable voyage on the sturdy schooner WANDERER.

Thor Heyerdahl, *Kon Tiki*. Heyerdahl believed (erroneously, as it turned out) that the Polynesians had originated in South America, and he completed a 3500-mile sea voyage on a balsawood raft to prove his theory.

Tristan Jones, *The Incredible Voyage*. The legendary sailor finds himself "a thousand times beyond the limits of endurance" on an intrepid six-year voyage during which he sails his small craft on both the Dead Sea and Lake Titicaca, the lowest and highest bodies of water in the world.

Sebastian Junger, *The Perfect Storm*. New England fishermen are caught in the biggest storm of the century.

Olivier de Kersauson, *The Sea Never Changes.* The narrative of de Kersausons 1988 solo round-the-world race on board Poulain, a 75-foot trimaran.

Robin Knox-Johnston, *A World of My Own: The Singlehanded, Non-stop Circumnavigation of the World in Suhaili.* The first non-stop singlehanded voyage around the world, for which the author won the Golden Globe trophy.

Alfred Lansing, *Endurance.* The incredible 800-mile Antarctic Ocean crossing by Shackleton and a handful of his men in an 18-foot whale-boat, as they go to seek help in South Georgia Island after their polar expedition fails. (Also read the account by Shackleton himself, listed below.)

Simon Leys, *The Wreck of the Batavia: A True Story.* The story of the mad heretic who led history's bloodiest mutiny off the west coast of Australia in the 18th century. (See also the book by Mike Dash, *Batavia's Graveyard.*)

David Lewis, *Ice Bird.* Lewis sets sail from Sydney, Australia in a small yacht, on a search for high adventure. He finds it in one of the least hospitable and most fascinating parts of the globe — the Antarctic.

Jack London, *The Cruise of the Snark.* London's account of the two years he and his wife spent sailing the South Pacific on their 55-ft ketch. His description of "surf-riding," which he dubbed "a royal sport," helped to popularize surfing in the U.S.

Bernard Moitessier, *The Long Way.* Around the world a time and a half without a landfall or encounter with another human being; his boat and the sea vibrate in unison. Also, *Tamata and the Alliance,* the epic story of his life.

Peter Nichols, *Sea Change.* Reflections on solitude and a lost love during an Atlantic crossing on a very old, very leaky sailboat.

Jonathan Raban, *Passage to Juneau.* A cruise up the inland passage along the Canadian west coast, as told by a writer with a splendid gift for describing man and nature.

Dougal Robertson, *Survive the Savage Sea.* A family survives weeks of high waves, shark attacks, and deprivation in a small dinghy and an inflatable rubber raft after a killer whale attack sinks their schooner.

William Robinson, *Deep Water Shoal.* The second (almost) solo around-the-world voyage (1932).

Tim Severin, *The Brendan Voyage.* The author tracks the discovery of America by sixth century Irish sailor saints, retracing the presumed route of Saint Brendan and his monkish crew in a replica of the leather boat they sailed in.

Sir Ernest Shackleton, *South: The Endurance Expedition.* The first attempt to cross the Antarctic continent, the ENDURANCE trapped in the ice and carried away, and the extraordinary voyage to South Georgia Island for help. (Also see the listing for Lansing, above).

James Simmons, *Castaways in paradise.* Incredible but true stories of castaways, each more extraordinary than the other.

Joshua Slocum, *Sailing Alone Around the World.* A retired sea captain makes the first solo around-the-world voyage. A must!

Robert Louis Stevenson, *In the South Seas.* Stevenson's account of his voyages to the Marquesas Islands and the Tuamotus.

Éric Tabarly, *Lonely victory, Atlantic race 1964.* Tabarly crosses the Atlantic in a record 27 days. And, in French, *Mémoires du large*, the great navigator's autobiography.

Paul Theroux, *The Happy Isles of Oceania.* A keen observer, one of the world's best travel writers, paddles his collapsible kayak through Polynesia and Melanesia.

Nicholas Thomas, *Cook: The Extraordinary Voyages of Captain James Cook* An extremely well-researched account of Cook's voyages.

Fiction

Joseph Conrad, *Typhoon.* A terrible storm wonderfully recounted by one of the greatest writers of the 20th century.

Daniel De Foe, *Robinson Crusoe.* Reread it as an adult.

William Golding, *Pincher Martin.* A castaway in desperate straits, by the author of *Lord of the Flies.*

Ernest Hemingway, *The Old Man and the Sea.* An old Cuban fisherman struggles to save his prize catch, but at the same time, a philosophical tale.

Homer, *The Odyssey.* The first great sea adventure in literary history. Ulysses's peregrinations over "the wine-dark sea" of antiquity. Try at least a few pages...

Jack London, *Tales of the South Seas.* Eight short stories depicting the havoc caused by the white man in the South Seas paradise. (Also see *The Cruise of the Snark* listed above.)

Pierre Loti, *Iceland Fisherman.* Life among the fishing communities of Brittany. A poignant tale by one of the finest descriptive writers of his day.

Herman Melville, *Billy Budd, Foretopman* (rather than *Moby Dick*, impossibly long — unless you're off on a two-year cruise). A superb short novel about a young British sailor in the mid 19th century who dares to strike an abusive officer.

Henri de Monfreid, *Secrets of the Red Sea*. Novel-narrative by the dashing French adventurer: derring-do, gun running and smuggling in the Red Sea in the first half of the 20th century.

Chris Nordhoff and James Norton Hall, *Mutiny on the Bounty*. The famous BOUNTY trilogy that served as a basis for several movies. (For historical authenticity, read Captain Bligh's own account, listed above, and *The Bounty: The True Story of the Mutiny on the Bounty* by Caroline Alexander.)

Robert Louis Stevenson, *Treasure Island*. Yo, ho, ho, and a bottle of rum! This classic adventure tale of pirates, sailing ships and buried treasure is a great read for adults as well as for younger readers. Or try Stevenson's *Kidnapped*.

Jules Verne, *20,000 Leagues Under the Sea*. Sail again with Captain Nemo aboard his extraordinary submarine, NAUTILUS. One of Verne's most exciting and successful *voyages imaginaires*.

References

[1] Allen, R.H., *Star Names: Their Lore and Meaning*, Dover, 1963.

[2] Aviso, website www-aviso.cnes.fr.

[3] Bascom, W., *Ocean waves*, Scientific American, Aug. 1959.

[4] Bekoff, M. and Byers, J.A., *Animal Play, Evolutionary, Comparative and Ecological Perspectives*, Cambridge Univ. Press, 1998.

[5] Bellwood, D.R., *Direct estimate of bioerosion by two parrotfish species, on the Great Barrier Reef*, Marine Biology, 121, p. 419, 1995.

[6] Bohren, C.F., *Clouds in a Glass of Beer*, John Wiley & Sons, 1987.

[7] Bolten, A.B. *et al.*, *Transatlantic developmental migrations of loggerhead sea turtles demonstrated by mtDNA sequence analysis*, Ecological Applications, 8(1), p. 1, 1998.

[8] Brownlee, D.C. and Ward, P.D., *Rare Earth: Why Complex Life Is Rare in the Universe*, Copernicus, 1999.

[9] Brusca, R.C. and Brusca, G.H., *Invertebrates*, Sinauer Ass., Inc., 1990.

[10] Butikov, E.I., *A dynamical picture of the oceanic tides*, American Journal of Physics, Vol. 70, No 9, p. 1001, September 2002, (www.ifmo.ru/butikov/).

[11] Campbell, N.A., and Reece, J.B., *Biology*, Benjamin Cummings, 2002.

[12] Chapelle, H.I., *The Baltimore Clipper, Its Origin and Development*, Tradition Press, 1965.

[13] Connor, R.C. and Micklethwhaite, P.D., *The Lives of Whales and Dolphins*, Henry Holt and Company Inc., 1984.

[14] Crane, N., *Mercator, The Man Who Mapped the Planet*, Henry Holt & Company, 2002, p. 204.

[15] Dor-Net,Z., *Columbus and the Age of Discovery*, William Morrow and Co. Inc., 1991.

[16] Doumas, C.G., *Thera: Pompeii of the Ancient Aegean*, Thames & Hudson, 1983.

[17] Druitt, T.H., and Francaviglia, V., *Caldera formation on Santorini and the physiography of the islands in the late Bronze Age*, Bulletin of Volcanology 54, p. 484, 1992.

[18] Ellis, R., *Monsters of the Sea*, The Lyons Press, 2001.

[19] Ellis, R., *The Search for the Giant Squid*, The Lyons Press, 1998.

[20] FAO, *The State of World Fisheries and Aquaculture*, United Nations Report, 2004.

[21] Florida Museum of Natural History, website: www.flmnh.ufl.edu/fish/Sharks/Statistics/statistics.htm.

[22] Fuson, R.H., *The Log of Christopher Columbus*, International Marine Publishing, 1992.

[23] Garrett, R., *The Symmetry of Sailing*, Sheridan House, 1987, p. 139.

[24] Gentry, Arvel, *Another look at the slot effet* and *More on the slot effect*, in *The Best of Sail Trim*, Sheridan House Inc., 2000.

[25] Giese G.S. and Chapman, D.C., *Coastal seiches*, Oceanus, Spring 1993, p. 38.

[26] Halstead, B.W., *Dangerous Marine Animals*, Cornwell Maritime Press, 1995.

[27] Heckstall-Smith, B., *Britannia and Her Contemporaries*, Methuen & Co, 1929, p. 6.

[28] Heiken, G. and McCoy, F. Jr., *Caldera development during the Minoan eruption, Thera, Cyclades, Greece*, J. Geophys. Res. 89, (B10), p. 8441, 1984.

[29] Hickman, C.P. and Roberts, L.S., *Animal Diversity*, WCB/McGraw-Hill, 1995.

[30] Hinz, E.R., *Understanding Sea Anchors and Drogues*, Cornell Maritime Press, 1987.

[31] Holm, D., *The Circumnavigators: Small Boat Voyagers of Modern Times*, Prentice Hall, 1974. Electronic version available at: www.mcallen.lib.tx.us/books/circumna/ci_a.htm.

[32] Hough, R., *Captain James Cook*, Hodder & Stoughton, London, 1994.

[33] International Marine Insurance Agency, website, January 2003.

[34] International Maritime Organization, *Reports on acts of piracy and armed robbery against ships*, Annual report for 2002, MSC.4/Circ.32, 2003.

[35] Irwin, G., *The Prehistoric Exploration and Colonisation of the Pacific*, Cambridge University Press, 1992.

[36] Judge, J., *The first landfall of Columbus*, National Geographic, p. 572, Nov. 1986.

[37] Kaufman, L. and Rock, I., *The moon illusion*, Scientific American, 207, p. 120, July 1962.

[38] Kirch, P.V., *On the Road of the Winds*, University of California Press, 2000, p. 100.

[39] Kurlansky, M., *Cod: A Biography of the Fish That Changed the World*, Walker & Co., 1997.

[40] Kusche, L., *The Bermuda Triangle Mystery — Solved*, Prometheus Books, 1975.

[41] László, V. and Woodman, R., *The Story of Sail*, Naval Institute Press, 1999.

[42] Lewis, D., *We, the Navigators*, University of Hawaii Press, 1972.

[43] Lohman, K.J. et al., *Regional magnetic fields as navigational markers for sea turtles*, Science, 294, p. 364, 2001.

[44] Lynch, D.K. and Livingston, W., *Color and Light in Nature*, Cambridge University Press, 1995.

[45] MacGregor, D.R. *Fast Sailing Ships: Their Design and Construction, 1775-1875*, Nautical Publishing Co., 1973

[46] Marchaj, C.A., *Seaworthiness, the Forgotten Factor*, Adlard Coles Nautical, 1996.

[47] Martin-Raget, G., *Les Plus Belles Escales à la Voile*, Hachette Livre-E/P/A, 1998.

[48] McNeill Alexander, R., *All-time giants: the largest animals and their problems*, Palaeontology, 41, p. 1231, 1998.

[49] Minnaert, M.G.J., *Light and Color in the Outdoors*, Springer, 1993.

[50] Mokady, O., Lazar, B., and Loya, Y., *Echinoid bioerosion as a major structuring force of Red Sea coral reefs*, Biol. Bulletin, 190, p. 367, 1996.

[51] Morison, S.,E., *The European Discovery of America, 1492-1616*, Oxford Univ. Press, 1974.

[52] Myers, R.,A. and Worm, B., *Rapid worldwide depletion of predatory fish communities*, Nature, Vol. 423, p. 280, 2003.

[53] NOAA, *Restless tides*, website http://co-ops.nos.noaa.gov/restles3.html.

[54] Oman, C., *In search of a cure for seasickness*, Cruising World, Safety at Sea, 1996.

[55] Oviedo, F., *La Historia General y Natural de las Indias*, Seville, 1535.

[56] Paccalet, Y. and Cousteau, J.-Y., *Cap Horn à la Turbovoile*, Flammarion, 1989.

[57] Paccalet, Y. and Cousteau, J.-Y., *À la Recherche de l'Atlantide*, Flammarion, 1981.

[58] Pacini, E., *La Marine, Arsenaux, Navires, Équipages, Navigation, Atterages et Combats*, 1844.

[59] Paxton, J.R. and Eschmeyer W.N., *Encyclopedia of Fishes*, Academic Press, 1995.

[60] Phillips-Birt, D., *An Eye for a Yacht*, A.S. Barnes and Company, 1955.

[61] Radhakrishnan, V., in D. Lewis, *We, the Navigators*, University of Hawaii Press, 1972, p. 311.

[62] Regenstein, J.M., Professor of Food Science, Cornell University, personal communication, 2004.

[63] Richardson, P.L., *Benjamin Franklin and Timothy Folger's first printed chart of the Gulf Stream*, Science, Vol. 207, p. 643, 1980.

[64] Romoli, K., *Balboa of Darien: Discoverer of the Pacific*, 1953.

[65] Schaefer, B.E., *The astrophysics of suntanning*, Sky & Telescope, p. 596, June 1988.

[66] Severin, T., *In Search of Robinson Crusoe*, Basic Books, 2002.

[67] Sfakiotakis, M., Lane, D.M. and Davies, J.B.C., *Review of fish swimming modes for aquatic locomotion*, IEEE J. Ocean Eng., Vol. 24, No. 2, 1999.

[68] Shane, S.H., Wells, R.S., and Orsig, B.W., *Ecology, behavior and social organization of the bottlenose dolphin: a review*, Marine Mammal Science 2(1), p. 34, 1986.

[69] Sherman, W.P., and Billing, J., *Darwinian gastronomy: Why we use spices*, Bioscience, Vol. 49 (6), p. 453, 1999.

[70] Shinn, E.A., et al., *African dust and the demise of Caribbean coral reefs*, Geophysical Research Letters, 27, p. 3029, 2000.

[71] Smith, F. and Brown, J.H., professors of biology, University of New Mexico, personal communication, 2004.

[72] Stocker, J.J., *Water Waves: The Mathematical Theory with Applications*, Interscience Publishers, 1957.

[73] Storandt, B., *The dangerous game of container roulette*, Cruising World, March 2002.

[74] Sulzberger, C.L., *Picture History of World War II*, American Heritage Society, 1966.

[75] Thomson, E.W., *Lightning and Sailboats*, Florida Sea Grant No. R/MI-10, 1992.

[76] Trehub, A., *The Cognitive Brain*, MIT Press, 1991.

[77] Triantafyllou, G.S. and Triantafyllou, M.S., *An efficient swimming machine*, Scientific American, March 1995, p. 40.

[78] United Nations, website: www.un.org/Depts/los/index.html.

[79] Van Dorn, W.G., *Oceanography and Seamanship*, Cornell Maritime Press, 2000.

[80] Wallace, W.J. *Oceanography — An Introduction*, Wadsworth Pub. Co., 1980.

[81] Waterman Talbot, H., *Animal Navigation*, Scientific American Library, Inc., New York, 1989.

[82] Weekley, E., *An Etymological Dictionary of Modern English*, Dover, 1967.

[83] Whidden, T., and Levitt, M., *The Art and Science of Sails*, St. Martin's Press, 1990.

[84] Whipple, A.B.C., *Restless Oceans*, Time Life Books, 1984.

[85] Wiegel, R.L., *Oceanographical Engineering*, Prentice Hall International, 1964, p. 277.

[86] Wilson, E.O., *The Diversity of Life*, Harvard University Press, 1992, p. 222.

[87] World Sailing Speed Record Council, website: www.sailspeedrecords.com.

[88] Zhang, J. et al., *Flexible filaments in a flowing soap film as a model for one-dimensional flags in a two-dimensional wind*, Nature, Vol. 408, p. 835, 2000.

Bibliography

The following is a short list of suggested reading covering some of the major topics discussed in this book.

Oceanography and marine life

Halstead, B.W., *Dangerous Marine Animals*, Cornwell Maritime Press, 1995. A good reference book on dangerous marine life.

Trefil, J., *A Scientist at the Seashore*, Charles Scribner's Sons, 1984. An excellent layperson's guide to the physical phenomena associated with the sea.

Van Dorn, W., *Oceanography and Seamanship*, Cornell Maritime Press, 2000. Physical oceanography from the recreational boater's perspective. A fundamental work, relatively technical but clear and accessible.

The sky

Meinel, A., *Sunsets, Twilights and Evening Skies*, Cambridge University Press, 1983.

Minnaert, M., *The Nature of Light & Color in the Open Air*, Springer, 1954. A classic work by a Dutch physicist enamored of the effects of light.

Weather

Bohren, C.F., *Clouds in a Glass of Beer*, Dover, 2001. Proof that the science of the everyday physical world can be fun.

Watts, A., *The Weather Handbook*, Sheridan House, 1999. A very clear explanation of weather by a practicing meteorologist.

Ships and their history

Greenhill, B., *The Evolution of the Sailing Ship — 1250-1580*, Conway Maritime Press, 1995.

Kemp, P., *The History of Ships*, Grange Books, 2002. An excellent overview of ships through the ages.

MacGregor, D. R., *Fast Sailing Ships: Their Design and Construction, 1775-1875* Nautical Pub. Co., 1973. Also by the same author: *Tea Clippers: Their History and Development, 1833-1875* and *British & American Clippers: A Comparison of Their Design, Construction and Performance in the 1850s.*

Lavery, B., *The Ship of the Line*, Conway Maritime Press, 1998.

The history of yachting

Clark, A., *The History of Yachting, 1600-1815*, Putnam's Sons, 1904. The fundamental reference work on the origins of the sport.

Giorgetti, F., *The History and Evolution of Sailing Yachts*, Chartwell Books, 2000. Yachting history as seen by a naval architect. Magnificent photos.

Herreshoff, L. F., *The Common Sense of Yacht Design*, The Rudder Publishing Co., 1946 and 1948.

Hold, D., *The Circumnavigators*, Prentice-Hall, 1974. The biographies of the solo — or almost solo — circumnavigators from Slocum to Moitessier. Out of print but available on the Internet at www.mcallen.lib.tx.us/books/circumna.

Pace, F., *Sparkman and Stephens*, Adlard Coles Nautical, 2002.

Robinson, W., *The Great American Yacht Designers*, Alfred A. Knopf, 1974.

Rousmaniere, J., *The Golden Pastime*, W.W. Norton & Co., 1986. A modern view of the history of yachting.

Stephens, O., *All This and Sailing Too*, Mystic Seaport, 1999. The autobiography of a true master.

Theory and design of sailboats

Bevan-Smith, J., *The Shape of Speed*, Reed New Zealand, 1999.

Brewer, T., *Ted Brewer Explains Yacht Design*, Camden International Marine, 1985. A layperson's guide by a naval architect.

Claughton, A.R., Wellicome, J.F., and Shenoi, R.A., *Sailing Yacht Design*, Prentice Hall, 1999.

Garrett, R., *The Symmetry of Sailing: The Physics of Sailing for Yachtsmen*, Sheridan House, 1987. A scientific approach to sailing.

Gutelle, P., *The Design of Sailing Yachts*, International Marine Publishing Co, 1979. An excellent review of the theory and practice of sailing yacht design.

Larson, L. and Eliason, R.E., *Principles of Yacht Design*, McGraw Hill, 1994. An essential book for anyone wishing to understand the details of naval architecture as applied to modern yachts.

Marchaj, C. A., *Aero-hydrodynamics of Sailing*, Adlard Coles, 1988. The Bible on the subject. His other works, *Sailing Theory and Practice* and *Seaworthiness: The Forgotten Factor*, are equally excellent.

Rousmaniere, J., *Desirable and Undesirable Characteristics of Offshore Yachts*, W.W. Norton & Co., 1987. The seaworthiness of pleasure boats analyzed by a committee of experts.

Skene, N., *Elements of Yacht Design*, Sheridan House, 2001. A reprinting of a classic work first published in 1904, but chock full of fundamental ideas and practical formulas that never go out of date.

Various authors, *The Best of Sail Trim*, Sheridan House, 2000.

Navigation

Lewis, D., *We, the Navigators*, The University Press of Hawaii, 1972. A fascinating explanation of the ancient Polynesians' navigation methods by a modern navigator who has put them into practice.

Index